LEARNING THE ART OF MATHEMATICAL MODELLING

ELLIS HORWOOD SERIES IN
MATHEMATICS AND ITS APPLICATIONS

Series Editor: Professor G. M. BELL, Chelsea College, University of London

Statistics and Operational Research

Editor: B. W. CONOLLY, Chelsea College, University of London

Baldock, G. R. & Bridgeman, T.	Mathematical Theory of Wave Motion
de Barra, G.	Measure Theory and Integration
Berry, J. S., Burghes, D. N., Huntley, I. D., James, D. J. G. & Moscardini, A. O.	
	Teaching and Applying Mathematical Modelling
Burghes, D. N. & Borrie, M.	Modelling with Differential Equations
Burghes, D. N. & Downs, A. M.	Modern Introduction to Classical Mechanics and Control
Burghes, D. N. & Graham, A.	Introduction to Control Theory, including Optimal Control
Burghes, D. N., Huntley, I. & McDonald, J.	Applying Mathematics
Burghes, D. N. & Wood, A. D.	Mathematical Models in the Social, Management and Life Sciences
Butkovskiy, A. G.	Green's Functions and Transfer Functions Handbook
Butkovskiy, A. G.	Structure of Distributed Systems
Chorlton, F.	Textbook of Dynamics, 2nd Edition
Chorlton, F.	Vector and Tensor Methods
Dunning-Davies, J.	Mathematical Methods for Mathematicians, Physical Scientists and Engineers
Eason, G., Coles, C. W. & Gettinby, G.	Mathematics and Statistics for the Bio-sciences
Exton, H.	Handbook of Hypergeometric Integrals
Exton, H.	Multiple Hypergeometric Functions and Applications
Exton, H.	q-Hypergeometric Functions and Applications
Faux, I. D. & Pratt, M. J.	Computational Geometry for Design and Manufacture
Firby, P. A. & Gardiner, C. F.	Surface Topology
Gardiner, C. F.	Modern Algebra
Gasson, P. C.	Geometry of Spatial Forms
Goodbody, A. M.	Cartesian Tensors
Goult, R. J.	Applied Linear Algebra
Graham, A.	Kronecker Products and Matrix Calculus: with Applications
Graham, A.	Matrix Theory and Applications for Engineers and Mathematicians
Griffel, D. H.	Applied Functional Analysis
Hanyga, A.	Mathematical Theory of Non-linear Elasticity
Hoskins, R. F.	Generalised Functions
Hunter, S. C.	Mechanics of Continuous Media, 2nd (Revised) Edition
Huntley, I. & Johnson, R. M.	Linear and Nonlinear Differential Equations
Jaswon, M. A. & Rose, M. A.	Crystal Symmetry: The Theory of Colour Crystallography
Johnson, R. M.	Linear Differential Equations and Difference Equations: A Systems Approach
Kim, K. H. & Roush, F. W.	Applied Abstract Algebra
Kosinski, W.	Field Singularities and Wave Analysis in Continuum Mechanics
Marichev, O. I.	Integral Transforms of Higher Transcendental Functions
Meek, B. L. & Fairthorne, S.	Using Computers
Muller-Pfeiffer, E.	Spectral Theory of Ordinary Differential Operators
Nonweiler, T. R. F.	Computational Mathematics: An Introduction to Numerical Analysis
Oldknow, A. & Smith, D.	Learning Mathematics with Micros
Ogden, R. W.	Non-linear Elastic Deformations
Rankin, R.	Modular Forms
Ratschek, H. & Rokne, Jon	Computer Methods for the Range of Functions
Scorer, R. S.	Environmental Aerodynamics
Smith, D. K.	Network Optimisation Practice: A Computational Guide
Srivastava, H. M. & Karlsson, P. W.	Multiple Gaussian Hypergeometric Series
Srivastava, H. M. & Manocha, H. L.	A Treatise on Generating Functions
Sweet, M. V.	Algebra, Geometry and Trigonometry for Science, Engineering and Mathematics Students
Temperley, H. N. V. & Trevena, D. H.	Liquids and Their Properties
Temperley, H. N. V.	Graph Theory and Applications
Thom, R.	Mathematical Models of Morphogenesis
Thomas, L. C.	Games Theory and Applications
Townend, M. Stewart	Mathematics in Sport
Twizell, E. H.	Computational Methods for Partial Differential Equations
Wheeler, R. F.	Rethinking Mathematical Concepts
Willmore, T. J.	Total Curvature in Riemannian Geometry
Willmore, T. J. & Hitchin, N.	Global Riemannian Geometry

LEARNING THE ART OF MATHEMATICAL MODELLING

MARK CROSS, B.Sc., Ph.D.
Head of School of Mathematics, Statistics and Computing
Thames Polytechnic

and

A. O. MOSCARDINI, M.Sc., Ph.D.
Senior Lecturer, Department of Mathematics and Computer Studies
Sunderland Polytechnic

ELLIS HORWOOD LIMITED
Publishers · Chichester

Halsted Press: a division of
JOHN WILEY & SONS
New York · Chichester · Brisbane · Toronto

First published in 1985 by
ELLIS HORWOOD LIMITED
Market Cross House, Cooper Street, Chichester, West Sussex, PO19 1EB, England

The publisher's colophon is reproduced from James Gillison's drawing of the ancient Market Cross, Chichester.

Distributors:

Australia, New Zealand, South-east Asia:
Jacaranda-Wiley Ltd., Jacaranda Press,
JOHN WILEY & SONS INC.,
G.P.O. Box 859, Brisbane, Queensland 40001, Australia

Canada:
JOHN WILEY & SONS CANADA LIMITED
22 Worcester Road, Rexdale, Ontario, Canada.

Europe, Africa:
JOHN WILEY & SONS LIMITED
Baffins Lane, Chichester, West Sussex, England.

North and South America and the rest of the world:
Halsted Press: a division of
JOHN WILEY & SONS
605 Third Avenue, New York, N.Y. 10016, U.S.A.

© 1985 M. Cross and A. O. Moscardini/Ellis Horwood Limited

British Library Cataloguing in Publication Data
Cross, Mark
Learning the art of mathematical modelling.
1. Mathematical models
I. Title   II. Moscardini, A. O.
511'.8   QA401

Library of Congress Card No. 84-29770

ISBN 0-85312-780-8 (Ellis Horwood Limited – Library Edn.)
ISBN 0-85312-850-2 (Ellis Horwood Limited – Student Edn.)
ISBN 0-470-20168-1 (Halsted Press – Library Edn.)
ISBN 0-470-20169-X (Halsted Press – Student Edn.)

Typeset by Ellis Horwood Limited
Printed in Great Britain by R.J. Acford, Chichester.

Table of Contents

Dedicated to our children

Matthew, Simon, Sarah
Danilo and Annamaria

Preface

Prior to the mid-1970s the emphasis of applicable mathematics in undergraduate courses was, principally, on the analysis of well-established mathematical representations of physical systems. Since the mathematical framework (i.e. the model) of these systems is well established, the effort was generally expended on special cases with analytical solutions or possibly on appropriate numerical analysis. Since the mid-1970s, however, there has been a growing interest in the problems associated with formulating the set of mathematical equations to represent a process or system. In other words, the emphasis is moving towards the development of mathematical models, rather than merely analysing the resulting set of equations, difficult though this may be. What are the reasons for this? In the decade from the mid-1960s to the mid-1970s the world changed dramatically for the graduate using mathematics in industry or commerce. Computer power had increased substantially whilst the cost decreased, computers became relatively easy to use and a wide variety of robust numerical techniques (and software) had become available. As these capabilities became apparent so the senior management of an ever-increasing number of industries became interested in using models to help in the analysis of complex processes or systems. So, a demand has since been growing for graduates who have acquired the skills of mathematical modelling together with some basic experience.

As mathematics, science and engineering graduates are now more likely to become involved in the development and application of mathematical models, it is useful if they are exposed to the ideas involved in the modelling of processes or systems during their under-graduate career. Having said this, expertise in mathematical modelling

is not something that can be easily taught, but it can be caught.
Therefore, it is important to provide an environment in which
students can acquire experience of mathematical modelling and learn
'the art'. Since the skills of modelling are caught rather than taught
then this, in many ways, separates it from the other more formal
branches of mathematics.

Since the skills of mathematical modelling are acquired through
experience, rather than by formal instruction, then it is a topic well
suited to individual study with an appropriate text. Therefore, the
primary aim of this book is to provide a framework which allows
the 'novice' modeller to 'catch' the ability to build mathematical
models to analyse real world problems arising from an industrial or
commercial context.

The structure and content of the book arises from a series of
courses in mathematical modelling, given by the authors at a number
of polytechnics and universities on both sides of the Atlantic. In
addition, therefore, to providing a structured introduction to mathematical modelling, chapters are included that would be specifically
useful to polytechnic and university teachers who are becoming
involved in this area. This extra material covers important aspects
of setting up and running a course, including how not to do so, and
how to assess student performance and utilise appropriate software.

The purpose of this book is not equip the student or the lecturer
with all that he requires, but rather to reveal something of what need
to be known. Mathematical modelling is experimental; expertise can
only be acquired through actually building models to analyse specific
systems or processes. No amount of reading can replace the need to
acquire some experience. This book merely aims to provide a framework which allows the student to gain some basic experience in
mathematical modelling.

The book is laid out in the following way. After an overview
of mathematical modelling, we examine how one particular model
was developed in some detail, just to put into context some of the
activities involved. Chapter 3 contains a salutary experience of one
academic who, whilst inexperienced in mathematical modelling
attempted to teach a course in the area (especially useful for
intending teachers).

Chapters 4 and 5 contain an approach to mathematical modelling
together with the scheme for either a self-learning or class-based
course. Chapter 6 contains the set of modelling scenarios that could

be used, whilst Chapter 7 contains a description of the type of computer software that should form an integral part of the students' experience. Chapter 8 concludes with an assessment of the utitity of the course and how we believe such developments should affect the structure, content and emphasis of mathematics degree programmes in general.

There are a large number of people who have influenced our thinking on this book and it is impossible to name them all. However, we would like to express our thanks to Jack Hobbs for many stimulating discussions (and arguments) in the early days, to Jim Caldwell for his collaboration (in contributing to one of the models), to Mary Thorpe for coding the software referred to in Chapter 7, and to Edie McFall for typing (and retyping) the manuscript. We are also grateful to our wives, Sue and Ruth, respectively, who have been both supportive of our efforts and rather patient during the writing of this book. The first draft was put together whilst the first author was a visiting professor at the Mineral Resources Research Centre, University of Minnesota, and we would like to acknowledge the facilities and assistance provided by the Director, Professor K. J. Reid.

Finally, we would like to acknowledge the contribution of David Peel who, though never an academic, actually provided much of the inspiration for the approach to learning mathematical modelling that has emerged in this book.

Mark Cross and Alfredo Moscardini
April 1984

CHAPTER 1

Mathematical modelling – an overview

1.1 INTRODUCTION

Most of us consider mathematics as consisting of a set of distinct branches that are both well compartmentalised and self-sufficient in the sense that they do not rely on other areas of science or engineering. This view is reinforced by the majority of problems we solve as students which tend to be well-defined with unique solutions. Each problem is complete in itself; it contains all the necessary information only and requires, at worst, a modification of a routine application of a mathematical technique to obtain the only correct answer. However, such a tidy state of affairs rarely manifests itself in real-life applications of mathematics. This is particularly true when we use mathematics to try and help understand the behaviour of some physical process or system. The system may be industrial, social, economic, or physicochemical, or involve a combination of these factors. In such cases, we are generally trying to answer an ill-posed question from an ill-informed politician, manager, engineer, etc., by finding a reasonable interpretation to a mass of mostly irrelevant data or information. In this context, the problem is to make sense of the question and determine a solution. When the process of problem-reduction involves transforming some idealised form of the real-world situation into mathematical terms, it goes under the generic name of mathematical modelling. Thus, mathematical modelling is an activity which requires rather more than the ability just to solve complex sets of equations, difficult though this may be.

Primarily, mathematical modelling utilises analogy to help understand the behaviour of complex systems. Analogy is used to help explain or understand unfamiliar situations all the time. For example, the phrase 'it tastes like ice cream' introduces a conceptual model of taste into our minds. In this sense, we continually use simplified representations and idealisations to enhance our understanding of systems and make predictions about their behaviour. So, just about every noun is a model of some event, process or idea – a point mooted by Plato [1] as long ago as 400 BC.

TREE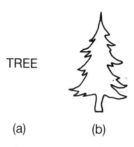

(a)   (b)

Fig. 1.1 – Model of a tree.

The word 'TREE' and the picture in Fig. 1.1 are both models of something in reality. They may not be very precise representations of a tree, but they do communicate something of relevance. Children model adulthood by playing mothers and fathers; medical students practise injections using oranges; pilots fly simulated missions; and expectant mothers practise nappy changing on dolls. Each of these activities involves some idealisation of reality, and models are in use. These models are not exact replicas of reality but contain only some of its essential elements. For example, unlike the model doll, young babies rarely lie still when having their nappies changed. In the same way, Newton's laws are really just 'idealised' models of the real physical world, as are all other existing theories of physical phenomena. So modelling is an activity which is fundamental to the scientific method. Each model merely reflects a mathematical description of a hypothesis concerning the behaviour of some physical process or other. At this point it is important to recognise that models must not be confused with reality. It is recorded historial fact that the lack of distinction between models and reality has often been instrumental in severely retarding progress.

It is paradoxical that initially successful models which enable quantum leaps in understanding to be made, often become the major stumbling-block to further progress because people are loath to discard the model even when it has been discredited. The classic example here is that of the medieval model of the solar system. The model of the planets describing circular paths with the earth as the common centre was successful in explaining night, day, the seasons etc. But there were several difficulties and facts that could not be explained. The Copernican solar system with elliptical orbits success-fully explained most of the outstanding problems but people simply refused to accept this model. The reason was that the early model also mirrored the medieval opinion of the place of man in the universe, and fulfilled the concept of perfection with its circular notions. This model was fixed into the faith and stability of the age. People were therefore loath to discard it, even though scientific facts disproved it. In this case, the existing model had become the established order of things and so began to obstruct rather than to stimulate and/or encourage progress [2].

1.2 MODELLING OBJECTIVES

So why build a model at all? After all, what can it tell us that we don't already know? To answer these questions it is important to realise that models never completely replicate a system and, therefore, they are not a unique representation and so can mean different things to different people. Consider how an accountant and a biologist view 'Bessie', the cow in Fig. 1.2. Their conceptual views of the same system or object are rather different since they are both heavily influenced by their own environment, background and objectives. The same is true when we come to the mathematical modelling of any system or process. Although, the motivation for building the model at all is usually the means to answering a particular question, the form of that question influences the way in which we build the mathematical model. The posed questions usually fall into one of five main categories:

  (i) system understanding,
 (ii) design,
(iii) optimisation with respect to prescribed constraints,
 (iv) control,
  (v) training.

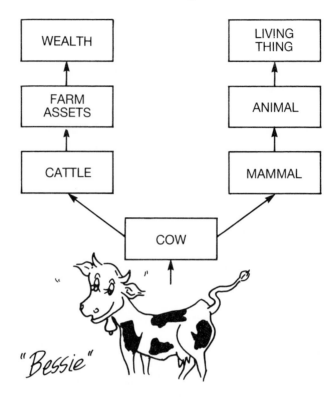

Fig. 1.2 – The ladder of abstraction.

Treating each of these in turn, the understanding afforded by being forced to rationalise one's conceptual view of a process or system and quantify the influence of each main factor, is often the single most important benefit from building a mathematical model. At best, if the model results match reality, it confirms the existing conceptual view; if they do not, then the modelling exercise highlights deficiencies in our existing conceptual understanding. It also affords the opportunity to refine and improve both our qualitative and our quantitative understanding of a particular system or process. An industrial example of this would be the rotary kiln process. Rotary kilns are used in a number of materials-processing operations to raise the temperature of the solids material to a specified (soak) level and often to ensure that certain chemical reactions also take place. The heat and mass transfer processes in the rotary kiln are illustrated in Fig. 1.3. The kiln axis is generally tilted at an angle to the horizontal

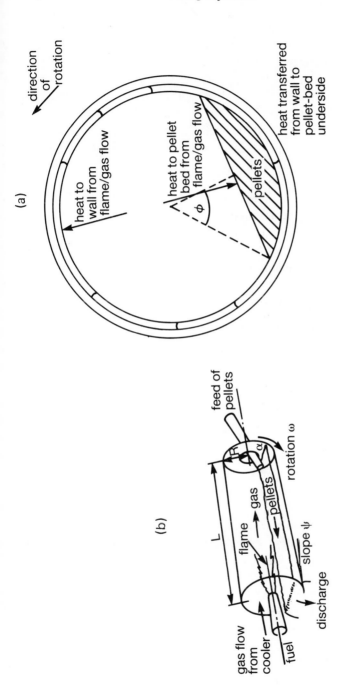

Fig. 1.3 — Conceptual views from rotary kiln. (a) Schematic diagram of rotary kiln. (b) Schematic diagram indicating major heat-flow paths.

and rotates at a few r.p.m. The particulate solids enter at the high end and tumble down to the exit. The gas enters at this point and flows countercurrent to the solid material. As illustrated in Fig. 1.3 (b) the heat exchange between the gas and the solids occurs by a number of routes. To further complicate matters a long combustion flame is sometimes used (for iron ore pellets, for example) and this distributes its heat along the length of the kiln. Although, simple in concept, it is difficult to intuitively estimate the fuel rates required to achieve a specified temperature, etc. and detailed mathematical models have proved useful in a number of industries.

In the design of new, larger or otherwise modified existing processes or systems, mathematical modelling has proved invaluable in a large number of industries. The conventional means of analysing a newly proposed design or operational procedure is to use a scaled down physical model. However, when a number of physical processes are involved simultaneously (e.g. heat and mass transfer), it is often impossible to devise appropriate scaling criteria. On the other hand, a mathematical model should be able to cope reasonably well with the concurrent influences of each major factor, provided each is built into the set of equations comprising the model. Another reason for using mathematical rather than physical models may be the expense of the latter. For example, wind tunnels have been used to evaluate the drag of proposed car designs for many years. However, it is expensive to experimentally identify sources of drag and modify the design to reduce it. A mathematical model, on the other hand, has the potential to identify sources of drag easily. Modifications to the design can be implemented and 'tested' quickly.

System or process optimisation is an important application of mathematical models in many areas today. The effects of optimising a system with respect to production rate, quality, cost, etc. may be assessed fairly readily. An example of system optimisation is in the design of fire-door arrangements to minimise the rate of spread of fires in buildings.

The use of mathematical models in system control has become widespread during the last decade or so. One example here involves the Treasury using economic models to assess how various budget options might best control the economy.

The use of models in training of personnel for high-risk tasks is another important application. Pilots have trained in aircraft simulators for over twenty years, and more recently the development of training simulators for advanced gas-cooled nuclear reactors attests to the importance of this motive for modelling.

1.3 MODEL CATEGORIES

There are two broad classifications of models — steady state and dynamic. Problems involving essential spatial complexity are usually modelled in the steady state initially and move to a dynamic form only if the required level of understanding cannot be obtained from the former. In many other problems the dynamic variation is the crux of the matter and, therefore, the modelling must cope with this aspect of the system's behaviour.

In further categorising 'mechanistic' models we tend to classify systems as continuous, discrete or a mixture. Mechanistic models are those whose mathematical description largely reflects a cause-and-effect sequence. By a continuous system we mean one that can be described by a set of ordinary and/or partial differential equations, such as flow through a river, etc. Discrete models involve the movement or variation of system measures which change in finite rather than infinitesimal ways. These models may involve a complex structure, but its state changes in a discrete way. An example of a discrete system is a vehicle production line.

If we have constructed a model by simply relating the output measures from a process to its input parameters via statistical equations then it is empirical in nature. Statistical or 'stochastic' models are often generated both in the early and latter stages of a model-building project. They are used early on to help establish some relationships between input parameters and output measures. Then, when a mechanistic model has been successfully developed, we may once again use a statistical model to condense its information into a more convenient form for use in fast optimisation or control.

In systems where it is very difficult to quantify the behaviour of some of its characteristics, fuzzy sets have been utilised. In this case the mechanistic structure is there, but the largely quantitative nature of the system is missing. The same can be true in the utilisation of graph theory to establish all the principal relationships within a system. In fact, an increasing number of mathematical techniques are being used as tools to model a growing number of systems.

1.4 A POTTED HISTORY

By and large, mathematical modelling is an art with a rational basis which requires the use of common sense at least as much as mathematical expertise. Eve [4] states that early mathematics has its origins in the more advanced forms of society in the ancient Orient, where it was perceived as a practical science to assist in agricultural and engineering endeavours. These activities required the computation of a usable calendar, a system of weights and measures, surveying methods for canal and reservoir construction, plus the evolution of financial and commercial practices for raising taxes and trade. In each case, careful quantification and prediction was the pre-eminent feature with the emphasis on arithmetic and mensuration.

In the last centuries of the second millenium BC new civilisations grew up to replace the rather static outlook of the ancient Orient. In a developing atmosphere of rationalism men began to wonder 'why', rather than simply 'how'. In this way demonstrative mathematics evolved and its deductive aspects came into being. With the ancient Greeks, mathematics became more abstract and was pursued for its own sake. However, other groups of workers, notably the Hindus, regarded mathematics as a tool for their astronomy. The use of mathematics in astronomical and optical applications has been dominant since the earliest times.

Whilst mathematical models have certainly been conceived since the origins of mathematics, it is worth considering developments since the seventeenth century a little more carefully. In the second half of that century Newton and Leibniz, working independently, developed the calculus. In the years that followed many other workers used calculus in the deliberate creation of mathematical models to represent and investigate natural phenomena [4]. Thus, the 'Genius Age', saw the development of many classical models of complex physical phenomena by great men. For example, the theories of gravitation by Newton, electromagentic waves by Clerk Maxwell and relativity by Einstein. The last is worth highlighting further. Space–time had been postulated by Minkowski, the idea of contraction was formulated by Lorenz and the existence of the ether had been contradicted by the Michelson–Morley experiment. The genius of Einstein was in a formulation that linked these previously unconnected ideas together into a relatively simple mathematical model we now know as special relativity. Einstein's ideas arose from posing himself the simple question, 'What would happen if I was travelling on a ray of light?'. Although the mathematics involved in general relativity became very complicated, Einstein was not one of the great mathematicians *per se*. Nevertheless, Einstein did make a tremendous contribution to our understanding of the universe by his modelling insight.

The 'Transitional Age' was heralded by the availability of various aids to arithmetic. Initially mechanical, and then electromechanical, desk calculators gave workers much greater freedom in the formulation of their models, since it was no longer necessary to ensure that analytic solutions were attainable. This was because the availability of desk calculators made numerical solutions relatively easy to obtain — at least for models with only a few equations.

The 'Contemporary Age' of mathematical modelling was not heralded by the invention of the electronic computer alone. The early stored program computers were scarce, difficult to use and rather slow. There are three major factors which have stimulated the rise to prominence of mathematical modelling over the last decade:

(i) the reduction of the man/machine interface communication barrier;

(ii) the development of a vast bank of practical numerical methods which yield accurate solutions to most consistent sets of equations;

(iii) the relatively small (and decreasing) cost of computing power.

The first of these has arisen as a result of relatively easy-to-use high-level languages and easier access to the machine hardware via terminals. The numerical techniques have amassed as a result of a worldwide effort in this area over the last twenty or so years. However, the single most important factor is the rapidly decreasing cost of computing power, to such an extent that now most children have been exposed to computing by the time they leave school. Paradoxically, this feature was largely encouraged by the extensive use of computers in the commercial world for a large volume and wide variety of exceedingly mundane tasks.

The above factors lead to adequate facilities being available for mere mortals to pursue the art of modelling in reasonable safety. The consequence of this has been a dramatic increase in mathematical modelling activity over the last ten years or so. Economic models now form the basis of many financial policies of advanced countries.

The ecological and environmental consequence of following particular development policies are assessed using models. Apart from these and a wide variety of models utilised in the 'soft' sciences, a vast development has taken place in the applied sciences and engineering fields covering every technological aspect of our society today. One may well pose the question, 'O.K., so it's quite feasible to build complex models, but why do so anyway?'. There are two main answers to this question and both are fairly obvious:

(i) For models, largely in science and engineering, which can be 'proved' (i.e. whose predictions can be substantiated against measurements as believable, reasonable and consistent), a relatively cheap and, often, quick way is provided for reinforcing system understanding, assessing the effects of system design, etc.

(ii) In many cases extensive 'proving' of models is impossible. Therefore, although their predictions must be treated with some care (or even a healthy scepticism), a model embodying the conceptual understanding of a system in quantifiable form, may still be the best way to help those with the responsibility for assessing the overall effects of a certain decision or design specification on a particular system. At the very least, it should provide an advance on existing rule-of-thumb schemes.

1.5 CONCLUSION

The purpose of this introductory chapter has been to indicate what mathematical modelling is, why we do it and how it has developed to its present state. We have established that mathematical modelling comprises an attempt to develop a coherent representation of the behaviour of some system or process, from a mass of largely irrelevant, disordered data. Furthermore, the motivation for modelling is that it provides a relatively cheap and rapid means of answering ill-posed questions concerning the system or process.

Thus since mathematical modelling is still developing throughout society it is vitally important that mathematicians, scientists and engineers be exposed to it. Not only should they be exposed to its utility, but they should also gain some basic experience and skill in using mathematics to provide quantified solutions to an increasing number of society's technical problems. This conviction is the motive for developing the material from which this book derives.

REFERENCES

[1] Plato, *Timaeus*, Penguin.
[2] A. Koestler, *The Sleepwalkers: Man's Changing Vision of the Universe*, Penguin.
[3] D. A. Peel, in *Modelling and Simulation in Practice*. (Eds. M. Cross *et al.*), Pentech Press (1979).
[4] H. Eve, *History of Mathematics*, 3rd edn, Holt Rinehart & Winston (1969).

CHAPTER 2

A real model exposed

2.1 INTRODUCTION

Before progressing to define how we should formally describe the process of mathematical modelling and certainly prior to discussing how to teach it to students, we felt it would be helpful to appreciate something of how a model of a real industrial process is developed. Often published technical papers which describe models only discuss the formulation that worked, and then in a well-structured organised way. The scaffolding which provided the support and motivation to develop in a given 'successful' direction has been taken away – leaving only a gleaming edifice.

The process to be examined here is the manufacture of 'green' or raw iron ore pellets prior to hardening in an induration machine and their subsequent charging as blast furnace burden. The reasons for using this process as an example are straightforward:

(i) one of us built the original model and therefore knows the *actual* development route;
(ii) the mathematics is fairly straightforward.

Before moving into the model development the next section contains a description of the pelletising process and summarises the data available at the time of the investigation together with both the relevant and irrelevant theories of pellet growth.

2.2 PELLETISING OF IRON ORE

The pelletising process essentially involves the agglomeration or sticking together of fine particulates into lumps which can be further

processed without blowing away as dust. The process, which was first
developed in the 1950s to utilise ores which were too fine for other
forms of treatment, has expanded rapidly so that by 1980 some 200
million tonnes were produced annually throughout the world.

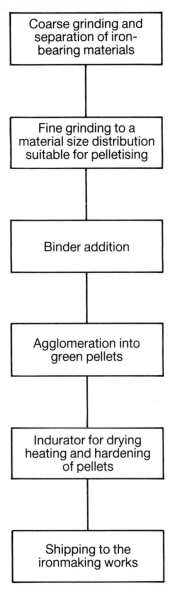

Fig. 2.1 – Unit operations involved in pelletising of iron ores.

Pelletising involves a number of 'unit' operations working in sequence. These are illustrated in Fig. 2.1. Having isolated the raw ore from the other elements found in the mined product, it must be ground to a suitable size distribution. It is then blended with a material additive to assist in the agglomeration process and transported to the pelletising circuits where the damp material is formed into raw or green pellets. Those green pellets which exceed a certain size move to the induration unit where they are dried and hardened by heating to 1300°C.

It is the expressed opinion of a large number of pelletising experts that the secret of successful operation of pellet plants lies in tight control of the pelletising circuits producing the green pellets. However, the problems of effective control of these circuits has hampered the efficiency of pellet production since its commercial inception over twenty-five years ago. Thus, the motivation for building a model of the process was to highlight the underlying reasons for the problems associated with the control of the pelletising circuits and to eliminate them, if possible. What follows is a relatively ordered summary of the recorded experience of a large number of pellet plant operators and research engineers. Initial information relating to the context of a modelling problem is rarely so conveniently described.

The process of green pelletisation consists of a simple feedback circuit (see Fig. 2.2), in which a rotating drum, whose axis is inclined at a small angle to the horizontal, is supplied with a flow of damp raw ore (with the binder, bentonite) at a constant rate plus the undersize material which has been returned from the screen. During the first couple of metres of the charge's progression through the drum the fresh feed material is taken up by the production of new seeds, after which the surviving pellets grow by a variety of mechanisms. At the discharge end of the drum the pellets are

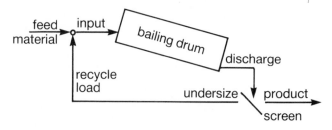

Fig. 2.2 – Flowsheet showing simple feedback structure of balling-drum operation.

creened, the oversize (greater than 9 mm at most current plants)
eing passed on as product to the induration system for drying and
iring, whilst the undersized are recycled to the charge end of the
rum where they participate in a further growth stage.

Both the recycle load and the output from these circuits are
ubject to a cyclic variation, normally referred to as 'surging'. In
ormal operation, the surge has a cycle time of 7–12 minutes
vhilst the average recycle load lies (typically) in the range 150–300%
f the feed rate and the recycle surge amplitude is about 20% of
his mean value. An example of this oscillatory behaviour which,
ncidently, has no deleterious effect on the consistency of product
juality when operating in a stable fashion, is shown in Fig. 2.3 (a).

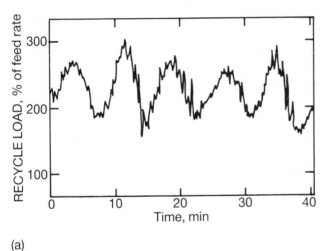

(a)

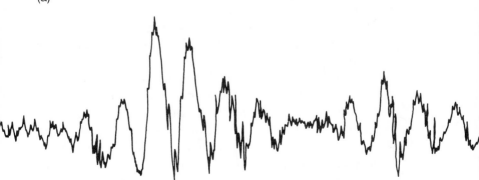

Fig. 2.3 – Variation of recycle load under (a) stable operating conditions and
(b) subject to operational disturbances.

These circuits are, however, liable to a phenomenon where the amplitude of the surge on both the recycle and output conveyor begins to vary dramatically, for no apparent reason. This behaviour which also causes the recycle load to increase up to +400% of the feed rate, is normally attributed to variation in one or more of the plant parameters (see Fig. 2.3(b)). It is detrimental to the control pelletising circuits – sometimes resulting in circuit breakdown – and also has deleterious effects on green pellet quality. Furthermore, it is not an uncommon occurrence for two drums in the same system (i.e. receiving the same material and nominally working under the same conditions) to be operating in a stable fashion whilst having different average recycle loads which could range from 150% to 300% of the input feed rate. It is also generally observed that high moisture contents result in a decreased recycle load, an elimination of the surge amplitude and an increase in the product size distribution. Bentonite tends to have a nullifying effect on the moisture and is often used to counteract the effects of excessive moisture in plant operation.

Some of the principal causes of the behaviour and the undesirable 'surging' described above are variations in the moisture content, bentonite level, size distribution of the feed material, rotational speed and the texture of the drum lining. Considerable effort has been expended over the last few years to eradicate these problems. However, variations still occur and the resulting loss of control – for periods up to an hour – is detrimental to the whole pelletising process in terms of both quality and quantity of output to the induration process.

Control, such as it is, is usually effected by a manual operator who adds moisture to damp out undesirable large oscillations. Although the operator generally has a moisture reading available to him he cannot rely on its accuracy. He must use other information to infer the actual moisture content and control the operation of the pelletising circuit. This information comes from two sources. The first is provided by a record of either the recycle or product load variation since, as stated above, these parameters are indicative of the moisture content. The second source is concerned with the observation that the surface appearance of the green pellets varies with moisture content (i.e. dull for low and shiny for high moisture content). With experience operators can learn to assess the moisture level directly by looking at the material in the drum, and adjusting the moisture accordingly.

Control of pelletising circuits is an art because the operator conventionally uses a combination of the recycle or product load and his judgement to adjust the water content in the drum, with two criteria in mind: (a) to make a product without excessive moisture and (b) to keep the recycle load and/or the surge amplitude from becoming too large. Furthermore, since acceptable pellet quality lies within quite a narrow range (see Fig. 2.4) excessive bentonite levels are often used to maintain quality. This practice is unsatisfactory for two reasons:

(i) excessive moisture in the product pellets to the induration process restricts its throughput;
(ii) bentonite is very expensive.

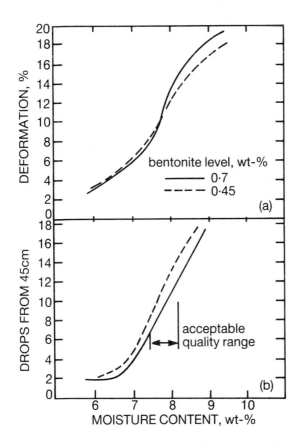

Fig. 2.4 – Variation of two green pellet quality measures with moisture.

So, if both the loss of control could be minimised and the circuits could be operated at optimum moisture and bentonite levels the effects would be beneficial in promoting the efficiency and production of high-quality pellets.

Finally, as a result of this account of operational experience, it is clear that the objective of the modelling exercise must be to highlight the factors that lead to the 'surging' problem and to find ways of ameliorating them.

2.3 MECHANISMS OF GREEN PELLET GROWTH

In the pelletising circuit there are only two groups of factors which modify the pellet size distribution. One involves the pellet generation and growth in the rotating drum and the other is the screen which classifies the distribution into product and recycle loads. The classification process is straightforward to simulate mathematically. The pellet generation and growth in the drum, however, depends upon its kinetic behaviour and, fortunately, the amount of published work describing this phenomenon is quite extensive. Although the natural forces responsible for the formation of agglomerates can result from a number of sources, it has been established that the most important contribution is made by capillary action between particles as a result of the air–liquid interfacial tension. Extensive laboratory studies using small batch pelletising drums (see Fig. 2.5) have identified three main phases of pellet growth.

  (i) *Nuclei phase* – particulate ore agglomerates initially into irregularly-shaped lumps, known as seeds, due to capillary action and other cohesive forces. These seeds have a highly porous moisture–air–material structure.

 (ii) *Transition phase* – the pellets enter this phase when, after further rolling, they are compacted so that the interstitial void volume decreases and the constricted capillaries fill with moisture. The pellet's surface becomes saturated so that it deforms easily and, as a consequence, growth continues quite rapidly, predominantly by a mechanism called coalescence (see Fig. 2.6(a)).

(iii) *Pellet growth phase* – Here the pellets are composed of a tightly packed interior surrounded by a thin wet shell and subsequent growth occurs by the mechanisms of coalescence, crushing and layering, and abrasion transfer, as illustrated in Fig. 2.6.

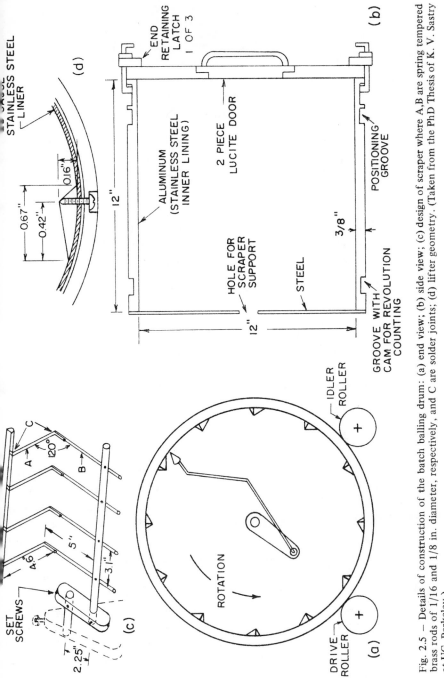

Fig. 2.5 — Details of construction of the batch balling drum: (a) end view; (b) side view; (c) design of scraper where A,B are spring tempered brass rods of 1/16 and 1/8 in. diameter, respectively, and C are solder joints; (d) lifter geometry. (Taken from the PhD Thesis of K. V. Sastry at UC, Berkeley.)

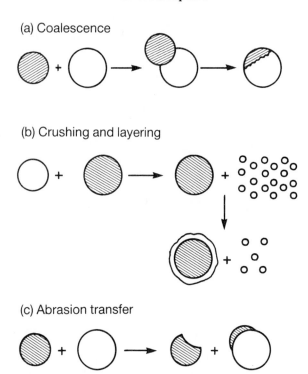

Fig. 2.6 – Representation of main pellet growth mechanisms.

Various groups of workers have attempted to demonstrate the dominance of one particular mechanism or other in pellet growth. Capes and Danckwerts [18,19] showed that crushing and layering was the dominant mechanism for (unnatural) closely sized sands. Furthermore, the green pellet size distribution in the batch pelletising drum was of a self-preserving nature. In fact, they also developed a theory of pellet growth based upon the crushing and layering mechanism. More precisely, they assumed growth took place by the crushing of the smallest pellet present and the redistribution of the material to the remaining pellets in proportion to their diameter. Their subsequent equation for the pellet size distribution was indeed found to be self-preserving. Some time later Kapur and Fuerstenau [17] performed a theoretical analysis based on the coalescence mechanism and again showed the pellet size distribution to be self-preserving in the same way as Capes and Danckwerts. More recently. Linkson and his colleagues [20] have shown that abrasion transfer is the important mechanism during the latter stages of pellet growth and once again confirmed the self-preserving nature of the pellet size distribution.

All the work referred to so far has been based upon results obtained from pelletising in batch, i.e. placing some material in the drum and observing the growth characteristics of the pellets as the drum rotates. Whilst this way reflects the situation for most of the time in commercial pelletising drums, it certainly does not do so at the feed end. Here pellets and fine feed material are mixed. To reflect this case Capes and Danckwerts also studied the effect on pellet growth of periodically adding fresh feed material to the drum. These experimental results showed that when fresh feed material was available both new seeds were produced and the existing pellets grew at a rate so as to maintain the same diameter distribution, i.e. growth was independent of pellet size.

A great deal of work has been published on investigations of the way in which variations in moisture content, bentonite level and particle size distribution of the raw material affects the rate of pellet growth. That the moisture content and particle size distribution should influence this factor is hardly surprising, since they are bound to affect the main cohesive, agglomerating force – capillary attraction. The effect of bentonite, which is added to the raw ore, to promote both faster drying rates and greater dry pellet strength during subsequent processing, is less obvious, affecting the growth rate by its ability to absorb moisture.

Sastry and Fuerstenau have examined in detail the effects on pellet growth rates of varying the moisture content and bentonite levels. Their results can be summarised in Fig. 2.7 which shows that increasing the bentonite content retards the pellet growth rate. The size distribution of the raw material also affects the growth kinetics; closely sized or coarsely ground material induces fast pellet growth whilst finely ground ores have the opposite effect.

Finally, apart from influencing pellet growth kinetics these factors (i.e. moisture content, bentonite level and size distribution) also affect the quality (i.e. strength and deformability) of green pellets. Both the wet and dry strength of green pellets are affected by a number of factors; in particular each strength measure increases with pellet size.

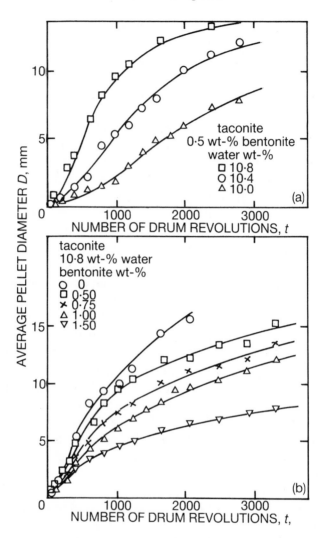

Fig. 2.7 – Effect of moisture and bentonite on green pellet growth rate (after Sastry and Fuerstenau [10, 11, 21–23]).

2.4 A CONCEPTUAL VIEW OF THE PROCESS

Essentially the green pelletisation process may be considered as one which modifies the pellet size distribution as it moves around the circuit. Ideally at the screen, the pellet size distribution below a specified size is retained within the circuit. In the pelletising drum the situation is quite complex with a large number of factors affecting the pellet size distribution. However, from the experimental evidence, it may be gleaned that if feed material is present, its layering onto existing pellets is the dominant mechanism. Hence, it seems worth hypothesising that there are two main growth regimes, distinguished by whether or not fresh feed material is available. It is also fairly clear that any new seeds must be generated in the regime where feed material is present. Thus, the first regime may be classified by the availability of feed material, the generation of new seeds and the growth of existing pellets by the layering of fresh feed only.

From the published material it would appear that pellet growth is very complex, with each growth mechanism exerting some influence. Some careful thought reveals that the fact that theoretical and experimental results produce self-preserving distribution for all mechanisms means it is not possible to ascribe pellet growth to one mechanism alone, and it is likely that all contribute to some extent. However, although it is impossible to distinguish between the individual growth mechanisms by mathematical analysis, it is interesting to note that all the theories make the tacit assumption that larger pellets grow at the expense of smaller ones. This assumption appears to be quite reasonable, bearing in mind the fact that green pellet strength is proportional to pellet size. Hence, it may no longer be so important to isolate the relative importance of the actual growth mechanisms, as to evaluate the overall rate at which growth occurs under operational conditions.

If the above reasoning is sound, then the evolution of the pellet size distribution in the second regime of the drum may be evaluated by assuming that the only effective pellet growth mechanism is larger pellets growing at the expense of smaller ones.

All the published laboratory results (see, for example, Fig. 2.7) show plots of the average pellet diameter or the cumulative distribution as they evolve with time. However, the difference in scale between laboratory and plant drums means that it is not possible to use directly the laboratory data in the modelling. From basic physical considerations, it seems reasonable that the two main factors which affect the rate of pellet growth are the number and force of collisions with other pellets. The number of collisions must be related to the distance rolled by pellets; the distance rolled by the pellets in the drum can be estimated with reasonable confidence. Many years ago, Saeman [25] showed that for lightly loaded drums, the average distance travelled per revolution is $4\pi R$, where R is the drum radius.

Hence, the distance travelled by the pellet charge in the laboratory drums may be estimated by

$$d_p = 4\pi R\, nT_r \qquad (2.1)$$

where n is the drum rotational speed and T_r is the residence time In commercial-sized pelletising drums, the residence time of the charge per pass may be estimated from

$$T_r = 0.037(\theta + 24)L/2\pi nRS \qquad (2.2)$$

where θ is the angle of repose, L is the drum length and S is the drum slope.

Recapitulating on our conceptual view of the process, it may be summarised as:

(1) The pellet growth in the drum may be divided into two regimes distinguished by the presence of fresh feed material.
(2) In the first regime seeds are generated and existing pellets grow by layering of fresh feed material only.
(3) In the second regime pellet growth proceeds by the smallest breaking down and its material being redistributed to the remaining pellets in proportion to their diameter.
(4) The rate at which pellets grow is related to the distance rolled by the pellets plus the force of the collisions between pellets.
(5) The screening operation is ideal in that only the fraction of the discharge pellet size distribution under a given size is recycled for further growth.

Having developed a clear conceptual view of the process, we will proceed to build a model which reflects that view. Of course, the conceptual view may be inadequate to describe the process suitably and this should be highlighted by the model. If this is the case, then our conceptual understanding must be refined and the model must be either modified or reformulated.

2.5 INITIAL ATTEMPTS AT MODELLING

This section contains the initial attempts at modelling the green pelletising process. A number of variants of this formulation were developed over a relatively short space of time and subsequently discarded — as was this approach.

2.5.1 Feed material present

Both feed material and recycled pellets are available on entry to the drum. Since the distance rolled by the pellets is equivalent to the time spent in the rotating drum, it may be assumed (from the results of Capes and Danckwerts [18, 19]) that pellet growth is directly proportional to the distance rolled. If y is the diameter of a pellet then

$$y(x) = y(0) + \alpha x$$

where x is the distance rolled in the drum and α is the growth rate parameter. The pellet size distribution may therefore be assumed to be a function of both the initial size as well as the distance rolled, i.e. it is assumed that $f(x, y)$ is the size distribution of recycled pellets with diameter y, after rolling a distance x.

As the recycled pellets grow, the volume of feed material diminishes; hence this variable may also be considered to be a function of distance rolled. Thus,

Volume of feed material present after pellets have rolled distance $(x + \Delta x)$	$=$	Volume of feed material present after pellets have rolled distance x	$-$	Amount used to generate new seeds	$-$	Amount used to augment growth of existing pellets

If it is assumed (i) that seeds are produced at a minimum diameter (δ) and (ii) that M are produced per unit volume of feed material per distance rolled, then in a distance rolled, Δx, the amount of feed material used to generate seeds will be

$$MV(x) \frac{\pi}{6} \delta^3$$

where $V(x)$ is the volume of fresh feed material available at distance x down the drum. In the same distance increment, each pellet will have increased in volume by

$$\frac{\pi}{6} \{[y + \alpha \Delta x]^3 - [y]^3\}$$

The material used to increase the volume of the whole size distribution is given by

$$\frac{\pi}{6} \int_0^\infty f(x, y)\{[y + \alpha \Delta x]^3 - [y]^3\}dy$$

Using the binomial theorem and assuming x is small, the above integral reduces to

$$\alpha \frac{\pi}{2} \Delta x \int_0^\infty f(x, y) y^2 \, dy$$

Substituting, the above equations into the diminishing volume balance on the feed material, gives

$$V(x + \Delta x) = V(x) - MV(x) \frac{\pi}{6} \delta^3 \Delta x - \alpha \frac{\pi}{2} \Delta x \int_0^\infty f(x, y) y^2 \, dy \tag{2.3}$$

which, in the limit, yields

$$\frac{dV}{dx} = -MV(x) \delta^3 \frac{\pi}{6} - \alpha \frac{\pi}{2} \int_0^\infty f(x, y) y^2 \, dy \tag{2.4}$$

When all the feed material has been used in pellet growth and seed generation, conservation of mass yields

$$\frac{\pi}{6} \int_0^\infty f(X, y) y^3 \, dy = V(0) + \frac{\pi}{6} \int_0^\infty f(0, y) y^3 \, dy \tag{2.5}$$

where X is the distance rolled at which $V(X) = 0$, i.e. the end of the first growth stage. From the above relations it should prove possible to evaluate the pellet size distribution $f(X, y)$, at the end of the first growth stage, from the input conditions only.

2.5.2 No feed material present

The model developed by Capes and Danckwerts which assumed that larger pellets grew at the expense of smaller, produced the following relationship between the pellet number and a given diameter:

$$\frac{n}{N} = \left(\frac{d_1 - d}{d_1 - d_s} \right)^\beta$$

where d_s and d_1 are the smallest and largest pellet diameter present, respectively, N is the total number of pellets, n is the number with diameter less than or equal to d and β is a positive constant. Thus,

$$d(x) = d_1(x) - (d_1(x) - d_s(x)) \left(\frac{n}{N} \right)^{1/\beta}$$

If, on entry to this regime, there are N pellets distributed as n_i with diameter $d_i(0)$ $(i = 1, \ldots, m)$ and it is assumed that the number of pellets decreases at an exponential rate then after rolling a distance X, there will be $N_1 = Ne^{-\gamma X}$ pellets on discharge. The exponential decrease in the total pellet size distribution is defined so that $d_1(0) > d_2(0) > \ldots > d_m(0)$ and after rolling a distance X only those greater than or equal to size $d_p(0)$ remain (where $d_p(0) > d_m(0)$) then

$$d_i(X) = d_1(X) - [d_1(X) - d_p(X)] \left(\frac{\sum_{i=1}^{p-1} n_{p-i}}{N} \right)^{1/\beta} \exp\left(-\frac{\gamma X}{\beta}\right)$$

In this formulation, the output diameters all depend on that of the largest value, $d_1(X)$. However, this may be evaluated from a combination of the above equation and volume balance of the material lost by disappearance of the smaller pellets and that gained by the remaining larger pellets, i.e.

$$\frac{\pi}{6} \sum_{i=p+1}^{m} n_i d_i(0)^3 = \frac{\pi}{2} \sum_{i=1}^{p} n_i d_i(X)^2 [d_i(X) - d_i(0)]$$

2.5.3 Comments on the model

Having formulated the model in this way initially, it was decided to first simulate the laboratory experiments. However, it soon become obvious that there were a number of problems associated with the implementation. The first was associated with the model of the regime where feed material was present. It proved impossible to find a suitable analytic expression for $f(x, y)$ and so it seemed natural to reformulate it in terms of a discrete size distribution as for the regime without feed material. Second, this latter regime required an input size distribution which was self-preserving in nature. The size distribution produced by the first growth regime was not self-preserving. Indeed, Capes and Danckwerts showed it would not be so. Thirdly, although experimenters had observed an exponential decay in the total number of pellets, this did not prove to be a very good way of modelling the laboratory measurements.

Finally, although the model could be 'tuned' to produce laboratory results reasonably well, the pellet size distribution entering the second regime in the pelletising circuits is composed of the undersize pellet size distribution and the new seeds that have been generated in the first regime. This distribution is far from being of a nature which can be adequately represented by the self-preserving form used in the second regime.

2.6 A PRACTICAL MODEL

There are a number of factors which inhibit the use of the model described above and some of its early variants. The eventual formulation that was used overcomes the problems highlighted above and is described below.

2.6.1 Feed material present

Suppose that on entry to the drum there is a volume V of feed material and N pellets from the recycle conveyor distributed as n_i with diameter d_i ($i = 1, \ldots, m$ and $d_1 > d_2 > \ldots > d_m$). The growth rate of the diameter of existing pellets is defined as Δ metres per metre rolled, and the production rate of seeds with diameter $d_{\min}$, as M per tonne of feed material per metre rolled. In fact, of course, seeds are produced with a size distribution but, in keeping with the simplicity of the current model, this is ignored.

The drum length is divided into a number of increments, then, at the end of the first, when the pellets have travelled a distance δ metres (say), the existing pellets have grown to a new diameter $(d_i + \delta\Delta)$ for ($i = 1, \ldots, m$) and there are now $n_{m+1} = MV\rho\delta$ new pellets (i.e. the seeds) with diameter, $d_{\min}$. The factor ρ is the density of the feed material. In other words, on entering the next increment there are $(N + MV\rho\delta)$ pellets distributed as n_i with diameter $(d_i + \delta\Delta)$ for ($i = 1, \ldots, m$) and $MV\rho\delta$ with diameter $d_{\min}$. Furthermore, the amount of fresh feed material in the next increment is reduced to

$$V^* = V - \delta\Delta \frac{\pi}{2} \sum_{i=1}^{m} n_i d_i^2 - \frac{\pi}{6} MV\rho\delta d_{\min}^3 \qquad (2.6)$$

Pellet growth is assumed to continue in this fashion until $V^* = 0$, at which point the growth moves into the next stage. In other words, comparing equations (2.4) and (2.6), the continuous formulation has essentially been replaced by a discrete representation.

2.6.2 No feed material present

On entry to this stage the feed material has been used up and the only pellet growth mechanism assumes the smallest pellet is crushed and the material redistributed to the surviving pellets in proportion to their diameter. In other words, the exit diameter for the surviving pellets is given by

$$d_i^{out} = d_i^{in} + \gamma (d_i^{in} - d_{k-1}) \, (i = 1, \ldots, k-1) \qquad (2.7)$$

where γ is a growth constant and $(k-1)$ designates the smallest class size of surviving pellets. The main problem is to evaluate γ and $(k-1)$. This may be done in the following way. The laboratory results illustrated in Fig. 2.7 show how the average pellet diameter varies with time (or, equivalently, distance rolled). In effect, these measurements provide the average pellet diameter as a function of the distance rolled, X, and effective moisture level, W, i.e.

$$d_{av} = f(X, W) \qquad (2.8)$$

Using this relationship, the average output diameter may be evaluated from the equation

$$d_{av}^{out} = f(X_0 + X_1, W) \qquad (2.9)$$

where X_1 is the distance rolled by the pellets in the second growth regime in the drum and X_0 is the distance defined by

$$d_{av}^{in} = \sum_{i=1}^{m} n_i d_i^{in}/N$$

The two unknowns $(k-1)$ and γ may then be evaluated from conditions which:

(i) define the already-known output average diameter, i.e.

$$d_{av}^{out} = \sum_{i=1}^{k-1} n_i d_i^{out} \left/ \sum_{i=1}^{k-1} n_i \right.$$

$$= (1+\gamma) \sum_{i=1}^{k-1} n_i d_i^{in} \left/ \sum_{i=1}^{k-1} - \gamma d_{k-1}^{in} \right. \qquad (2.10)$$

(ii) evaluate the material transfer due to the crushing and redistribution procedure,

$$\frac{\pi}{6} \sum_{j=k}^{m} n_j d_j^{\text{in } 3} = \gamma \frac{\pi}{2} \sum_{i=1}^{k-1} n_i (d_i^{\text{in}} - d_{k-1}^{\text{in}}) d_i^{\text{in } 2}$$

i.e.

$$\gamma = \frac{1}{3} \sum_{j=k}^{m} n_j d_j^{\text{in } 3} \bigg/ \sum_{i=1}^{k-1} n_i d_i^{\text{in } 2} (d_i^{\text{in}} - d_{k-1}^{\text{in}}) \qquad (2.11)$$

The solution of these equations provides a drum discharge pellet distribution with n_i pellets of diameter

$$d_i^{\text{in}} + \gamma (d_i^{\text{in}} - d_{k-1}^{\text{in}}) \text{ for } i = 1, \ldots, k-1$$

To recapitulate:

- from an initial pellet size distribution and mass flow rate of both the pellets and fresh feed material,
- the model will predict the pellet size distribution at the exit from the drum,
- the influence of moisture and bentonite on growth rate is reflected by equation (2.9) and the parameters (M, δ) whilst that of drum dimensions, etc. is reflected by equations (2.1) and (2.2),
- the unknowns to be solved for come down to γ (a growth rate measure) and k the smallest pellet class size remaining at the end of the drum, as represented by equations (2.10) and (2.11).

This formulation avoids the constraint of specifying the form of the pellet size distribution and the precise way in which the pellet numbers decrease.

The model is completed by simulating a screening operation on the discharge distribution, i.e. passing on those over a specified diameter as product and recycling the remainder which then forms the charged pellet size distribution.

Finally, the data used in these calculations is mainly derived from the results of Sastry and Fuerstenau [10, 11, 12–23] with the dependence of the (M, Δ) growth characteristics on the effective moisture content being given by

$$M = 1000(1707W - 16061) \text{ tonne of feed material/m rolled}$$
$$\Delta = 0.001(0.005W - 0.046825) \text{ m/m rolled}$$

Also, since the force of the collisions will be somewhat greater than in the laboratory drum, the growth rate is 'scaled' to reflect this factor.

2.7 MODEL RESULTS AND DISCUSSION

In discussing the model's predictions it is worth bearing in mind the original objective for building the model, that is, to highlight the factors that lead to the 'surging' problem and to find ways of ameliorating them.

A large number of model runs were made under a variety of constant operating conditions. From these results it became clear that the model would predict 'surging' type behaviour of regular oscillations in both the product and recycle loads, no matter what practical operating conditions were used. A typical set of computed results is shown in Fig. 2.8. It was also found that although the period and amplitude of the oscillations may be either enhanced or damped, depending upon the operating conditions (see Fig. 2.9), they

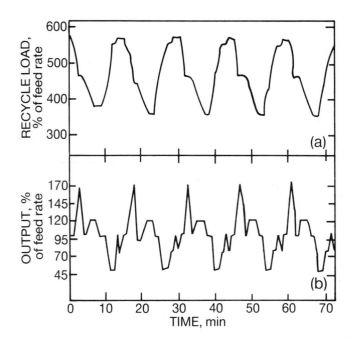

Fig. 2.8 – Model prediction of recycle load and corresponding output variation when operating in a stable fashion.

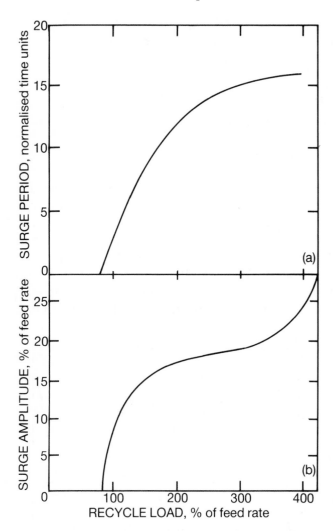

Fig. 2.9 – Model results of surge characteristics as function of recycle load.

were always present. Furthermore, Fig. 2.10 shows that the model predicts the average recycle load increasing as either the moisture decreased or the bentonite increased. These results were encouraging because they tallied with the recorded experience of operators of pelletising circuits.

The model was then used to study the problems likely to be experienced when trying to control the process. For one reason or

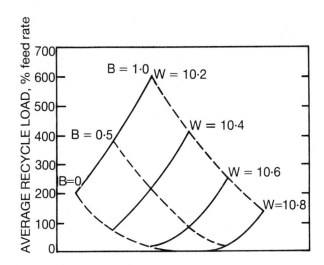

Fig. 2.10 – Model prediction of average recycle load as function of moisture content W(wt-%) and bentonite level B(wt-%).

another, the moisture content of the incoming fresh feed material can vary by about 2% by weight. Figure 2.11 shows an example of a transient response of the model to a step change in the moisture content. A whole series of results involving dynamic variations in the operating conditions were generated to investigate the problem of process control. Combined with the results assuming constant

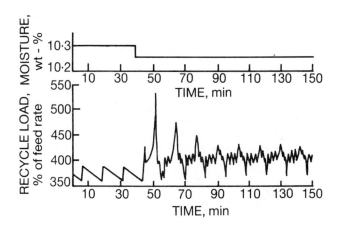

Fig. 2.11 – Model predictions of pelletising circuits response to a small decrease in moisture content of the material in the drum.

operating conditions, three main factors may be highlighted as complicating the task of control:

(a) the relatively large change in the stable operating point of the average recycle load and product size for small changes in the operating conditions;

(b) the complications induced by a superimposed oscillation on both the recycle and product loads which, in addition, is also very sensitive to changes in the operating conditions, and

(c) the variation in the response of the system to changes in the operating conditions.

In other words, since the measure used by the operator to control the operation is unreliable for the first few cycles after a variation, the task of effective control becomes very difficult. This was demonstrated by designing an automatic controller and testing it out on the model. It only performed as well as a diligent 'manual' operator and suffered from all the same shortcomings.

At the same time as the modelling work was being developed a control analysis of the pelletising circuit was carried out which confirmed that surging was, in fact, a limit cycle oscillation. This result combined with the model predictions substantiated that the surging behaviour was inherent in the basic design of the pelletising circuit. Since it is clear that surging is the source of all control problems associated with the pelletising circuit, the model was then used to assess the utility of a number of circuit design modifications (which were suggested by the control analysis) in improving process control.

The first scheme which was analysed was the pellet hold-up device (see Fig. 2.12). The idea behind this scheme was for the device

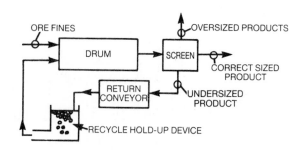

Fig. 2.12 – Schematic of recycle hold-up surge controller.

(a hopper) to accept all the undersize returns and supply the recycle input to the drum at a controlled rate, thus smoothing out the fluctuations in the recycle load. This scheme was simulated using the model. A typical result is shown in Fig. 2.13 from which it is clear that although the oscillation is damped a very large hopper would be required to do so effectively. Since the green pellets are fairly fragile and could be subject to severe pressure in the hopper, the ideal is probably not practically feasible for iron ore pelletising at this time.

A second scheme which was assessed was that of partial feedback (see Fig. 2.14). Here a small fraction of the recycle is taken off the recycle load and recrushed. This was simulated using the model and the result from Fig. 2.15 indicates this scheme may well be feasible since there is a relatively small amount to be siphoned off the recycled load. Fig. 2.16 serves to emphasise this point by illustrating the response of the recycle load to a small step change in the moisture content when the recirculated fraction is 90%.

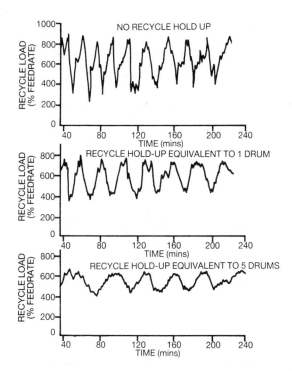

Fig. 2.13 – Typical responses for the recycle hold-up surge controller.

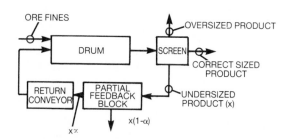

Fig. 2.14 – Schematic of partial feedback surge controller (whereby only a fraction α of the undersized product is recycled).

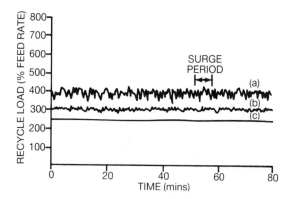

Fig. 2.15 – Variation of the recycle, (a) under standard conditions, (b) when 95% of the undersize is recirculated, and (c) when 88% of the undersize is recirculated.

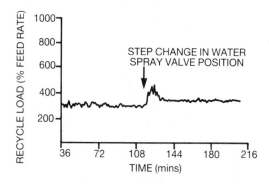

Fig. 2.16 – Recycle-load response to a step change in water sprays, with partial feedback ($\alpha = 0.9$).

The final scheme assessed was that of tandem operation. Here various fractions from the recycle conveyor of adjacent drum circuits are supplied to the neighbouring drum. The motivation for the scheme was to find a convenient way of utilising the partial feedback idea without discarding any material. The scheme design is shown in Fig. 2.17 along with a typical result (Fig. 2.18). Essentially, the scheme does not work because the circuit 'gain' of the tandem circuit is not less than that of two single circuits. In other words, the surge problem is not suppressed at all but merely shifted around the tandem circuit. Thus at the present time the scheme which looks most promising is the partial feedback.

2.8 CONCLUSIONS

The purpose of this chapter has been to provide a real example of how mathematical models are developed (or evolved) for industrial applications. The model is not elegant, nor does it contain any high level or 'new' mathematics. Put simply, familiar mathematics has been used as a tool to provide a pragmatic quantitative description of a real industrial process. It is pragmatic in the sense that it is just complex enough to represent the essential features of the system relevant to the particular areas of interest. The model described in this chapter is neither the only one that could be developed for the pelletising circuit nor is it necessarily the best. It is simply

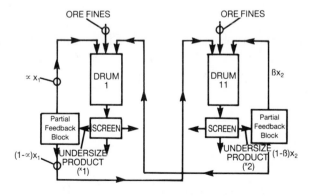

Fig. 2.17 – Schematic of tandem-drum surge controller.

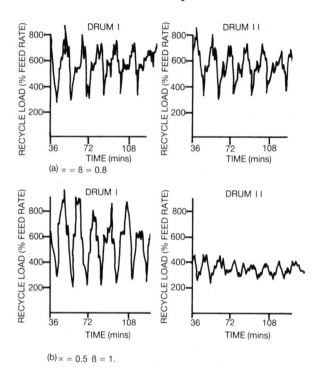

Fig. 2.18 – Typical simulation results for tandem operation. (a) $\alpha = \beta = 0.8$.
(b) $\alpha = 0.5$ $\beta = 1$.

good enough to be able to answer those questions which originally motivated its development.

Finally, this chapter also demonstrates the relative importance of mathematics in model-building. Its importance is only incidental, i.e. the formulation is not so mathematically or numerically complex that a solution cannot be adequately generated. On the contrary, the emphasis is place upon trying to understand the behaviour of a system or process.

REFERENCES

(a) General

[1] D. F. Ball *et al.*, *Agglomeration of Iron Ores*, Heinemann (1973)
[2] K. V. Sastry (Ed.), *Agglomeration 77*, AIME (1977).

[3] P. Somasundaran (Ed.), *Fine Particles Processing*, AIME (1979).

[4] O. Molerus and W. Hufragel (Eds), *Proc. 3rd Int. Symposium on Agglomeration*, Nurembourg (1981).

[5] C. E. Capes, *Particle Size Enlargement*, Elsevier (1980).

[6] P. J. Sherrington and R. Oliver, *Granulation*, Heyden (1981).

[7] B. Waldie, *Size Control in Particle Growth Processes*, Institution of Chemical Engineers (1983).

[8] P. C. Kapur, 'Balling and granulation', in *Advances in Chemical Engineering* (1979).

(b) Special

[9] D. W. Fuerstenau *et al.*, *CIM Bulletin*, **69**, 67 (1976).

[10] K. V. Sastry and D. W. Fuerstenau, *Trans. AIME*, **258**, 335 (1975).

[11] K. V. Sastry and D. W. Fuerstenau, *Powder Tech.*, **7**, 97 (1973).

[12] C. E. Capes *et al.*, SME-AIME Preprint No. 75–B–25.

[13] H. Rumpf, in *Agglomeration* (Ed. W. Knepper), Interscience (1962).

[14] M. Tigerschiold and P. A. Ilmoni, *Proc. AIME Blast Furn. Conf.*, **9**, 18 (1950).

[15] D. M. Newitt and J. M. Conway-Jones, *Trans. Inst. Chem. Engrs.*, **36**, 422 (1958).

[16] P. C. Kapur and D. W. Fuerstenau, *Trans. AIME*, **229**, 348 (1964).

[17] P. C. Kapur and D. W. Fuerstenau, *Ind. Eng. Chem. Process Des. Develop.* **8**, 56 (1969).

[18] C. E. Capes and P. V. Danckwerts, *Trans. Instn. Chem. Engrs.*, **43**, 116, (1965).

[19] C. E. Capes, and P. V. Danckwerts, *Trans. Instn. Chem. Engrs.*, **43**, 125 (1965).

[20] P. B. Linkson *et al.*, *Trans. Instn. Chem. Engrs.*, **51**, 251 (1973).

[21] K. V. Sastry and D. W. Fuerstenau, *Ind. Eng. Chem. Fund.*, **9**, 145 (1970).

[22] K. V. Sastry and D. W. Fuerstenau, *Proc. Inst. Brig. Agglom.*, **12**, 113 (1971).

[23] K. V. Sastry and D. W. Fuerstenau, *Trans. AIME*, **252**, 254 (1972).

[24] R. A. Bayard, *Chem. Met. Eng.*, 100 (March 1945).

[25] W. C. Saeman, *Chem. Met. Eng.*, 508 (October 1951).

(c) Papers arising from the work

[26] M. Cross, *Ironmaking & Steelmaking*, **4**, 159 (1977).

[27] M. Cross and P. E. Wellstead. *Simulation,* **30**, 55 (1978).
[28] P. E. Wellstead *et al., Int. J. Min. Proc.,* **5**, 45 (1978).
[29] P. E. Wellstead *et al., Automation in Mining, Mineral and Metals Processing,* IFAC (1978).
[30] P. E. Wellstead *et al., Trans. Inst. M. C.,* **2**, 86 (1980).

CHAPTER 3

Teaching mathematical modelling – a salutary experience

3.1 INTRODUCTORY REMARKS

Most books and papers on the art of mathematical modelling extol its virtues, but say very little about how it can be taught successfully. However, before being able to understand which particular teaching or learning methods are successful and which others are not it is helpful – if not a little humbling – to experience some degree of failure. The unfortunate thing about failing as a teacher is that it can affect the future of so many others. So we tend to try to minimise such occurrences. Nevertheless, it is helpful to experience schemes which fail to work entirely satisfactorily and this chapter summarises an early attempt of one of the authors at teaching modelling to one group of students.

The particular group of students concerned had a rather poor background in mathematics; they were graduating in a topic which featured the interaction between society and engineering. The modelling course was given during the students' second year. In the first year they had had a series of lectures in basic mathematics involving mainly elementary calculus and statistical methods.

The course had commenced with a couple of lectures on the principles of, and a systematic approach to, mathematical modelling. This featured some detailed discussion of a modelling structure which indicated how the process of modelling should proceed from a real world description of the system to a suitable mathematical model. The structure used was the Open University seven-box method and is illustrated in Fig. 3.1. At the conclusion of this discussion the

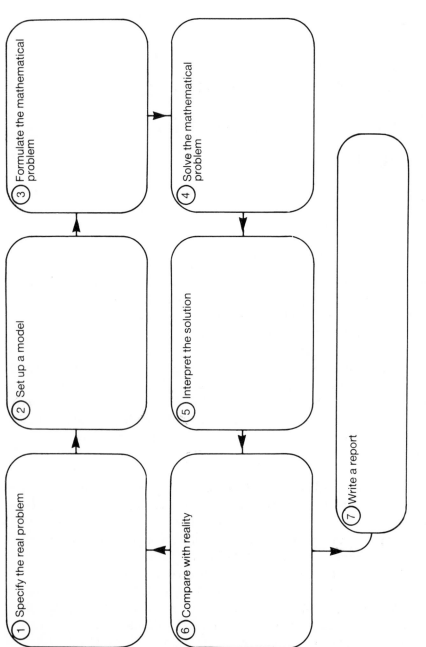

Fig. 3.1 — The mathematical modelling process as illustrated by The Open

students were separated into groups and set to work on the potato problem. This is a simple problem used by the Open University to illustrate the process of mathematical modelling. Essentially, it involves building a model to decide whether it is better to buy small or large potatoes in order to maximise the yield of material suitable for human consumption per unit cost. During this exercise the students went through some of the basic stages involved in modelling. They came up with criteria and proposed a means by which to decide how to maximise the yield per unit cost.

Although the students worked well enough through the potato problem, they complained somewhat about its fictitious nature and expressed an interest in solving a more realistic problem.

3.2 THE MAIN PROBLEM

Week 3

After some careful thought by the teacher the class was set the following problem: 'Does the number of cash points (or checkout stands) affect the profit of a large supermarket? If so, can this number be optimised?'.

This problem was selected for the following reasons:

(a) it was a realistic situation with which the students would be familiar;

(b) the problem was not well-defined and there were no obvious quick solutions; thus

(c) the students would naturally be led to a modelling situation;

(d) the complexity of the problem would lead to idealisation to render it tractable;

(e) the need for a systematic approach to modelling would soon become apparent;

(f) the problem could be tackled with the students' existing level of mathematical knowledge.

In fact, the above statement (f) is not quite true since some queueing theory is needed in the later stages of the project. However, the relevant formulae could be used without necessarily understanding their derivation. Actually, the teacher felt that even this aspect was important, because more and more, students without an extensive background in mathematics will be required to utilise the predictions of models in professional life to evaluate political, social, economic and design policies for a whole range of situations. In fact one of the major objectives of this course was to prepare the students for this professional activity.

3.3 STUDENT PERFORMANCE

In the first meeting the problem was presented and a class discussion invoked. As the students began to introduce large numbers of factors that could affect the system under consideration, it was stressed to the class that they would have to both make assumptions to simplify the real system and collect some data from a real supermarket. No clear direction as to how to proceed was given to the students since the teacher was interested in how they would respond 'unprompted' The class was divided into four groups of six students each, and excused formal (relevant) classes for the next week to gather data and work together.

Week 4, Week 5

At the beginning of the fifth week the student groups reported on their findings:

(a) Group 1 – From their observations and discussions they high-lighted the following facts:

   (i) the service rate at till was fairly constant (i.e. 2 min/person)
   (ii) supermarket was located in the town square;
   (iii) square floor area;
   (iv) everyone used a trolly (or cart);
   (v) maximum queue length of four.

   Furthermore, they observed that on the average 20 people entered the store per minute, the average queue time was 2 minutes and so a queue length of 4 people meant 5 cash desks (i.e. no. of cash desks = rate at which people enter store/average queue length).

   From their presentation the teacher deduced that only half the group had contributed anything at all. However, it was clear that the group had made some progress which could be developed.

(b) Group 2 – This group achieved virtually nothing. Two of its members had deduced the following formula,

No. of cash desks =

$$\frac{\text{average number of people in store}}{\text{length of average queue} \times \text{average time in store}}$$

but had failed to discuss the problem to any extent or to visit a store to observe it working and collect some basic data. A poor response in the time available.

(c) Group 3 – This group produced nothing at all! They claimed that they had no idea how to start the problem.

(d) Group 4 – This group had obviously put in some work. They had established an aim, viz. to make departure and arrival rates the same. They had also assumed that if x people live in a catchment area containing y supermarkets then x/y will go to each.

A visit to a store had yielded the following data:

(i) till service takes 31 s per person;
(ii) on average, one person leaves the store every 2–5 s and arrives every 2–6 s;
(iii) the average person spends 5.5 min in the store.

Having acquired this data, however, they were not too sure how to use it!

With regard to an overall assessment, it was strongly suspected that only about 9 out of the 24 students took any real active role. Overall, enthusiasm was low and no-one really had any idea how to construct a model beyond the trivial level. Almost all the students felt that the problem was too large and difficult for them.

At this stage the class was informed that the teacher was less than pleased with their response, bearing in mind the amount of work that most of them had put in. They were instructed to go and make a further attempt, and to hand in an individual report each the following week.

Week 6

At the beginning of the next week the nine active students handed in reports which represented quite reasonable attempts at modelling the supermarket problem. The rest of the class either could not or simply failed to produce a written report on their work – if indeed they actually done any.

At this stage the teacher felt it would be useful to develop a fairly simplistic model with the class. The following basic model was then presented.

3.4 TEACHER'S BASIC MODEL

At the outset it was pointed out that although the following model was not the most sophisticated it represented an attempt to get to grips with the problem.

As an initial attempt it is probably worth considering one service point in the supermarket in isolation. The situation is illustrated in Fig. 3.2. Some customer arrival and service patterns have to be assumed whilst the output from the model which is of interest may be summarised as:

(i) queue length;
(ii) total time spent in the queue by the customer;
(iii) proportion of the time for which the server is idle.

Finally, it may be worth imposing a constraint on the maximum number of people in a queue (e.g. if this value is 6 then a seventh person would be discouraged and not join).

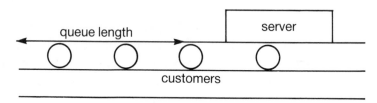

Fig. 3.2 – Illustration of the simple customer queue-server system.

Assume a customer arrives at the end of the queue every a seconds (i.e. has an arrival rate of $1/a$ per second) and leaves the service point every b seconds (i.e. has a departure rate of $1/b$ per second). If it is also assumed that there are r customers in the queue to start with then:

(i) $a = b$ implies customers arrive and leave at the same rate, so that the queue length always remains at r and each customer has to wait br seconds to be served. So, provided the initial queue length is acceptable (i.e. not too large), only one till is needed.

ii) $b < a$ implies that customers leave faster than they arrive and so eventually the queue size will diminish to zero or one. It is easy to show that an initial queue length of r customers takes $(rab)/(a - b)$ seconds to disappear and it will obviously be uneconomic to use more than one service point unless the value of r is unacceptably high. It is interesting to note that once the queue has disappeared, then the cashier is idle for $100(a - b)/a\%$ of the time.

ii) $a < b$ implies that customers arrive faster than they leave, so that unless any constraints on its size are imposed, the queue will grow indefinitely from an initial length r to $rt(b - a)/ab$ after t seconds. In fact, this model can be used to provide an indication of how many service points are required to alleviate the queue growth problems. For example, two service points essentially halves the service time and, therefore, the time interval, b, between customers leaving. This means that n service points reduces the interdeparture time to b/n. When this is the same or less than a, the queueing problems will disappear. A similar treatment could be developed to cover a maximum queue length at a single service point.

.5 TEACHER'S COMPLEX MODEL

'eek 7

lthough, the simple model has permitted some insight into the roblem at hand, it does have some major approximations with erhaps the most significant being the assumption of a constant rival rate. There are two ways of coping with this feature:

ι) perform a computer simulation using random number generators, etc.;

ɔ) fit some probability distribution to the arrival rate.

he first alternative was considered too difficult for the current oup of students, so the second was followed. A Poisson distribution as selected; this had the advantage that the students had dealt ith this topic in their earlier statistics course.

Examine the problem at the busiest time of each day. If z people rive every five minutes then e^{-z}, ze^{-z}, $\dfrac{z^2}{2}e^{-z}$ etc., give the probability that 0, 1, 2, etc., people will join the queue within the time. Furthermore, if x is the number of people served in a 5-minute interval then the probability that a customer has to queue is z/x. Thus, the probability that a customer will be served immediately $(1 - z/x)$, the average number in the queue is $z^2/x(x - z)$, the average time a customer spends in a queue is $5z/x(x - t)$ and the average time taken to serve a customer is $5/x$. Obviously, the above formulae assume that $z < x$ otherwise an infinite length queue would grow.

At this stage the students were given some simple exercises with these formulae so that they might get a 'feel' as to what the equations actually tell them.

Having spent some time on this the next stage was to extend the model to cover multiple service points. The queueing theory for this level of generality was complex and rather beyond the mathematical ability of the current group of students. However, this should not have prevented them from being able to use complex formulae. As such, statistical formulae for several service units were presented to the students with a general explanation of their meaning. A couple of students, who were interested in programming, then volunteered to write a computer program to evaluate the formulae for a variety of arrival distributions, service times and service points.

3.6 RECAPITULATION

Week 8

Although the class was well into the problem by this stage, the teacher concluded that it had been a mistake to become so involved in the queueing aspect of the problem whilst neglecting the whole picture somewhat. Even though the multiservice unit formulae were adequate and the computer program had been well written, the complex mathematics had alienated most of the class who had begun to lose interest once more. It was now time to stand back and examine the problem again from a wider perspective.

After some thought it was agreed that what we were trying to do was: 'Examine and evaluate the effect of the number of cash check out points on the profitability of a supermarket'. Accordingly the goals we were trying to satisfy amounted to determining the optimum number of checkout points that will:

(i) minimise customer inconvenience at peak periods, and
ii) utilise a minimum number of staff,

ɔ that the profitability of the store will be maximised by providing
n economic and efficient service.

To achieve these goals it was considered important to estimate

(i) queue waiting time,
ii) queue length,
ii) economic cost of cash registers.

inally, the store takings should be maximised.

The model used to estimate the weekly profit was:

weekly profit (P) = fixed % of total takings (T)
                   − customer disgruntlement cost (C)
                   − cash till running costs (R)

e. $P = T - C - R$.

Through consultation with a local supermarket manager it was
ɛvealed that a medium-size store operates on a net profit of $4-5\%$
nd since the total weekly takings amount to £40,000–£50,000
ıen $T \simeq £2,500$. Each till costs about £5,000 and has a life span of
bout 20 years. This yields a weekly running cost for each till of £5.
hus, for K tills the weekly running cost of all the tills, $R = £5K$.

It is, needless to say, difficult to estimate the cost of customer
issatisfaction to the supermarket. There are obviously a number of
ıctors that affect the degree of customer satisfaction, with:

(i) store prices,
ii) availability of commodities,
ii) degree of crowding in the store, and
v) time in checkout queues

eing perhaps the most important.

Without a detailed data collection exercise it is difficult to
ıantify the effects of the first three factors above. However, by
ssuming a Poisson distribution for both arrival and service lines
vith means λ and μ) for all K units operating in parallel, the mean
me each customer spends in a queue, t_K, can be estimated. If a
ıstomer does not object to spending t_0 minutes in a queue for the
ısh till, then a measure of customer dissatisfaction is afforded by
$\alpha(t_K - t_0)$.

At this stage we now had some sort of model of the system. However, in order to obtain some quantitative measures of the effects of dissatisfaction, a discussion on a suitable questionnaire was instigated. This led to a series of questions to ellicit the expected queue time, how long customers would wait in a queue before not coming again and whether they would prefer cheaper goods or no queues. The students were then dispatched to conduct the survey and amass the required data.

Week 10
Only half of the students came to the class and, as a result, there was not enough data to obtain any quantitative conclusions from the survey. So the time was used to find out what the students thought of the course they had followed. The main objections of those who attended this session were:

(i) the problem was too vague,
(ii) there was not enough data available,
(iii) the exercise was not 'real' mathematics,
(iv) the problem was poorly posed because the survey result indicated that the major factor affecting a store's profitability was the cost of its goods.

3.7 TEACHER'S GENERAL COMMENTS

'For the majority of the students the exercise was not successful. Interest quickly flagged as the students began to appreciate the possible complexity of the problem and their enthusiasm was never re-kindled. Most found the problem too vague and continually complained of not knowing how to proceed with its solution. Furthermore, any data that became available was never in the form they would have liked. As a consequence, many of the students 'opted out' and it was, therefore, very difficult to get them to actually produce any work at all.

'In answer to the students' complaints and criticisms, I would respond that their situation was consciously created in order to force them to appreciate the need for a systematic approach to obtaining some form of solution to the problem. Furthermore, I felt that the problem as stated reflected the vagueness likely to be met in professional life. With regard to giving a lead, I was in a delicate position since I felt that any suggestions I made would immediately be taken as fact and pursued to the bitter end by the class.

'Although, one or two students did make concerted efforts to develop a model the rest failed to produce anything of any real consequence. Furthermore, most of the last few classes generated very little in the way of student participation, and so degenerated into 'chalk-and-talk' lectures. The most lively class involved the discussion on the desirability of a survey and the necessary questionnaire. However, it produce nothing of any real consequence.

'Overall the course provided a sobering experience for me and in the future I would ensure that in any such exercise:

(i) Group sizes would be restricted to a maximum of three in order to ensure that everyone contributes to the project.

(ii) Avoid using surveys and questionnaires as this detracts from the main object of the exercise.

(iii) Make much more data available.

(iv) Instil the need for a systematic approach in the students at a much earlier stage.'

CHAPTER 4

The art of mathematical modelling I – the process of mathematical modelling

4.1 INTRODUCTION

Mathematical modelling is an art because it involves not only the development of a set of skills but also experience and intuition. In fact, it is fair to say that intuition and its results (e.g. insights, pragmatic descriptions of a system, etc.) are the factors that distinguish good modellers from mediocre ones. Modelling, therefore, requires much more than just the ability to solve a set of equations – no matter how complex they may be.

Because experience and intuition form such a significant role in the practice of modelling it is difficult to provide an adequate formal description or definition which makes any practical sense. A number of books provide a relatively brief discussion based upon a block diagram of the form illustrated in Fig. 3.1. However, this type of discussion completely understates the expertise required to build 'good' mathematical models, Moreover, modelling can also appear as a 'trivial' or soft option to classical mathematicians. This is because the mathematical content only constitutes part of a wider exercise that is extremely challenging intellectually.

Of all the discussions on the process of mathematical modelling the one that best gives intuition its primary role is by Peel [1]. For many of those involved in the practice of mathematical modelling as a profession, it is this description that most accords with their experience. Accordingly, it is the approach developed by Peel that is described below.

4.2 THE PROCESS OF MATHEMATICAL MODELLING

The chain of activities which should take place as a problem is solved may be conveniently divided into four main segments or stages. These stages and their relation to the real-world problem and its context are illustrated in Fig. 4.1. Broadly speaking, problem solution moves through these stages sequentially, although obviously one may revert to earlier stages of the process in the light of experience. Notice here that the discussion is around 'problem solution' rather than model building *per se*. This is simply because, in practice, models are developed to assist in the solution of real-world problems. Therefore the success of a model is governed by how much it contributes to a reasonable solution to the problem at hand.

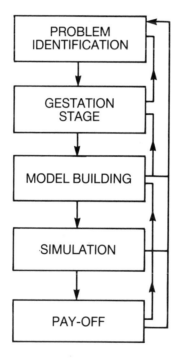

Fig. 4.1 – The basic stages of problem solving using mathematical models.

When mathematical modelling is used to aid in the solution of a problem, any activity prior to the formulation of a model is designated the *gestation* stage, whilst those which follow the decisions reached as a result of the simulation stage constitute the *pay-off* stage. The distinction between the *model building* and *simulation* stages may not be so obvious, but it is, nevertheless, an important one. Both of these stages are of a hypothetico–deductive nature and the distinction between them is one of form and content. In the model building stage the hypotheses are concerned with the mechanistic structure of the system to be modelled, whilst in the simulation stage these are taken for granted (at least, temporarily) and the concern is largely with the effect of variations in parameter values (i.e. operation of the model).

4.3 THE GESTATION STAGE

Consisting as it does of those activities which occur before starting to formulate a model, the gestation stage is somewhat nebulous. This oft-neglected stage is very important; it involves the gathering of background information, acquiring an understanding of the system under consideration, sorting out relevant data from the mass available and generally becoming familiar with the whole context of the problem. In particular it is important to recognise that:

   (i) a problem exists;
   (ii) a solution is possible.

The recognition of a relevant problem is both culture- and time-dependent (e.g., no-one worried about the ethics of nuclear weapons in the 1914–18 war). Furthermore, the recognition of a problem is strongly influenced by related problems that have been solved in the recent past. By contrast, the ability to recognise that a solution is possible is dependent on the individual. It is at this stage that it becomes important to identify both the dominant and the limiting factors in a problem, to discard much of the irrelevant information and to acquire a sound conceptual understanding of what the problem really is and what has to be overcome to solve it. It is during this phase that modellers should 'see the light' or experience 'Eureka'.

Intuition and, therefore, experience play an important role at this stage. These factors coupled with the identification of red herrings and ability to encompass or restrict problems to the relevant aspects only, influences the kind of model that will be produced. Intuition arises largely as a function of experience and, as such, cannot be taught. Such attributes have to be encouraged by the framework of any course in mathematical modelling.

On departing from this stage the modeller should have a good conceptual understanding of the problem and its context. This degree of understanding should have produced an unambiguous question to be answered, some objectives to be satisfied and a means by which to achieve them. In the process a substantial amount of the available data has probably been discarded and a conceptual framework evolved.

4.4 THE MODEL BUILDING STAGE

Having devised an acceptable conceptual framework and a suitable data subset the *formulation* of the mathematical model is often fairly straightforward. Having exposed the important mechanisms and contributory factors, each is considered in turn and described mathematically to some extent. Depending on what has already been previously written and the degree of detail required, the sub-model of each mechanism may be complex or consist of a simple empirical equation. When each main contributory mechanism has been included it is important to ensure that the resulting set of equations is consistent and that their solution has a reasonable chance of falling within the realms of reality (e.g., negative mass is not usually appreciated).

After formulating a consistent set of equations a *solution* method must be devised. This may be analytical, although for most realistic systems it is much more likely to involve some numerical evaluation. There is a possibility that the construction of a suitable solution method may involve the development of numerical techniques. However, this requirement is becoming increasingly rare as the number of robust automatic computer packages becomes available.

Any set of equations is a model of something; the question that needs to be answered here amounts to: is your set of equations a reasonable model of the system under consideration? In other words, the *validation* of your model is a vitally important aspect of model building. By the time the above question is first asked the initial set of equations constituting the model have been developed, a solution technique both developed and implemented, and some initial results obtained. If the system being modelled already exists then some measurements may be available to characterise its behaviour. If not,

then usually some laboratory-scale measurements will have bee
made or a qualitative expectation of the system's behaviour will b
available. In these cases, the basic questions to be answered include:

(a) Does the model match available data?
(b) Is it within the physical constraints of the system?

Another problem to be aware of at this stage is that of scale. Much c
the validation work may be performed with bench-scale data. It i
important to ensure that some attempt has been made to include th
influences of scale within the model equations.

If the model is found wanting with regard to some aspects of it
validation, then it is likely that one or more of the mechanisms wa
not accounted for as precisely as it should have been. This is remedie
by modelling this aspect of the system in a little more detail – suc
modifications may involve the inclusion of empirical or more comple:
physical equations.

If the model fits the data well in a qualitative way it can then b
'tuned' by varying some of the parameter values to improve th
quantitative fit of the model results to the data. Another useful wa
to *refine* the model is to non-dimensionalise the equations and the:
examine which terms contribute most to the model performanc
This approach provides a means to eliminate terms that really have
negligible effect on the overall performance of the model and henc
the system. It is important, however, to avoid removing the 'physic;
nature' of the equations until it is confirmed that the model appear
to work reasonably well.

4.5 THE SIMULATION STAGE

The building of the model having been completed, it may then b
used in anger, at least until further improvements become necessary
Proposals for modifications to the real-world situation are cor
verted into specific parameter sets and the model is then used t
run an experiment, i.e. to assess what would happen in a hithert
unexperienced situation. A set of simulation runs should be fairl
carefully designed to observe the effects of parameter changes anc
hence, identify which are the most critical with respect to bot
design and control.

It is still important even at this stage to carefully evaluate th
whole set of model predictions before placing your faith in then

Peel encourages modellers to be sceptical of their results and not to believe them in any absolute sense. It is not unusual to have model shortcomings highlighted at this stage which require some further reformulation of the model. When the results start to arrive in force and things 'look good' it is sobering to recall the asymmetry between proof and disproof of a model. Whilst the simulation results may support the hypotheses made in constructing the model, they cannot prove exhaustively that they are true. What makes matters worse is that just one awkward set of results can directly prove that your model is inadequate or wrong. However, disproving a model is at least as valuable as substantiating it, since this discredits any misconception of the mechanisms responsible for the system's behaviour.

Finally, when the model reaches the stage that simulation runs are to be carried out on a regular basis, it then becomes important to make the computer program as efficient as possible with respect to both machine processing time and ease of use by others.

4.6 THE PAY-OFF STAGE

Many scientists and engineers involved in modelling tend to abandon the problem once they have generated a set of consistent results which contain answers to the original questions that were posed. However, although this final stage is often neglected, indeed, sometimes not even recognised, it is just as essential a part of mathematical modelling as are the earlier stages. After all, there seems little point in expending a great deal of effort and nervous energy in developing the model plus a significant cost in computing resources in simulating a system if no-one is going to bother with the implications of its results.

Having first built the model, generated a series of model results and assessed them carefully, the modeller should have an excellent comprehension of the system under consideration. He should therefore be in a strong position to make balanced recommendations to the client (i.e. the one who posed the original problem) regarding operation, design and/or control of the system.

Not only should the modeller make recommendations, but he should also be prepared to actively sell them to his client. The reason for this is that, unless he is fortunate, the modeller will be just one of a number of inputs to the decision making procedure. So it is important to ensure that the major model recommendations are taken cognisance of along with financial constraints, etc. But it is also important to watch out for the *flip* reaction. This is the process whereby the client (often a busy senior manager) either disbelieves or dismisses everything to do with the model and its development, until he sees model results that behave like the real system. Then he 'flips' and believes everything produced by the model. This is a potentially dangerous time because the modeller's attitude must now change from selling to holding back, ensuring that the model is not used for any purpose for which it is not relevant.

Finally, having built the model, made recommendations and then observed either the construction or modification of the system it is useful for the modeller to make an appraisal of just how well the model performed. It is at this stage that one can learn most about the modelling process itself. During the model's development mistakes are bound to have been made, though one hopes they are only small ones. When appraising the model's performance one does so from the advantage of knowing what actually happens. This allows one to see and appreciate any conceptual or even computational errors that lingered in the model. It is in this assessment that a modeller grows in experience.

4.7 SOME HINTS

At this early stage it is useful to note the following set of warnings that should help novice and experienced modellers alike to build useful models.

(a) Remember, all models are partial, they never represent the entire system.
(b) Models are usually initially built for a specific purpose; be careful, therefore, of using them for purposes other than originally intended.
(c) Do not fall in love with your model.
(d) Take your model results with a grain of salt unless directly validated in some way.
(e) Do not distort reality to fit the model.
(f) Do not retain discredited models.
(g) Do not extrapolate beyond the region of validity of your model hypotheses.
(h) Keep ever-present the distinction between models and reality.

(i) Be flexible and willing to modify your model as the need arises.
(j) Keep your objectives ever present by continually asking yourself, 'What am I trying to do?'.

This set of hints is by no means exhaustive, but it should help novice modellers to avoid some of the more obvious pitfalls associated with the construction of mathematical models.

4.8 CONCLUSION

At this point it is worth noting that, as a paragraph based upon the block diagram in Fig. 3.1 is an understatement of the expertise required to build useful mathematical models, so too is the more detailed description of the process of modelling presented in this chapter. The above description merely represents a more detailed summary of what modelling involves, an indication of its constituent parts and their interaction. Although, it is by no means complete, it certainly demonstrates that modelling is not a trivial process. As such, the skills required to be able to develop mathematical models take time to acquire, It is the acquisition of these skills that constitutes the motive for the next chapter on how to teach mathematical modelling.

REFERENCE

[1] D. A. Peel, in *Modelling and Simulation in Practice* (Eds M. Cross *et al.*), Pentech Press (1979).

CHAPTER 5

The art of mathematical modelling II – an approach to the teaching of mathematical modelling

5.1 INTRODUCTION

So far we have discussed mathematical modelling from a number of viewpoints. These have included: why its current level of importance to our society is so great; what modelling involves; a description of the development of a real model; and the recollections of a sobering experience in trying to teach mathematical modelling. The main purpose of this chapter is to explain our philosophical approach to the teaching of mathematical modelling skills and to describe a teaching scheme that has been used and that 'worked' reasonably well in some non-quantifiable sense. However, before plunging into the 'nitty-gritty', it is worth spending some time considering the problems of re-orientation that have to be faced by both teachers and students when they become involved in mathematical modelling

5.2 REORIENTATION

The current structure of most mathematics courses, whether for specialists or for scientists, engineers and others, is still rather formal It is precise, well-structured and in many cases largely concerned with rigour. This background contrasts starkly with many of those aspects of the real world which have to be modelled. So a significant proportion of students suffer a severe 'culture shock' when meeting mathematical modelling as a specific discipline for the first time

Students find modelling disturbing for a variety of reasons, with perhaps the most important being:

(i) the open-ended nature of the problems,
(ii) the emphasis on providing some insight into a system rather than on elegant mathematics,
iii) the lack of a 'unique' solution,
iv) the necessity of combining a number of mathematical techniques rather than one obviously 'applicable' method,
(v) coming to terms with the fact that (even quasi-) real-life problems can rarely be solved in thirty minutes or so.

Nor are the problems associated with modelling confined to students only — teachers also have a difficult task. In fact, teachers have a two-fold problem. The first involves their own background and much of their academic experience. Here they have been involved in inculculating a disciplined formal approach to mathematical concepts with the objectives of exposing the generalities of mathematics· on the one hand, and the preciseness of its nature on the other. When building pragmatic mathematical models this approach is cast aside and by conventional mathematical standards — anything goes! So before teaching students the teacher first has to reorient himself towards a new conceptual view of mathematics — one where mathematics is sublimated to the more general needs of solving a real world problem rather than just a set of mathematical equations. Having reoriented himself the teacher then has to overcome the natural inertia of students to open-ended problems by providing enough motivation for the student to generate the confidence to start building mathematical models. Therefore, the systems to be modelled by the students must be:

(i) such that the context of the problem can be understood reasonably quickly and without requiring any specialist knowledge that the student does not already possess;
(ii) realistic enough to be interesting, yet not so complex that the student cannot formulate a model without using mathematics beyond the limits of his present knowledge;
iii) such that the resulting set of equations can be solved fairly readily by the student.

These criteria place severe constraints on the type of systems which can be modelled, especially for students at early stages in their college careers. Paradoxically, it is imperative that, as soon as possible, students should both be aware of the general open-ended nature of real-world problems and be able to master the skills required to formulate consistent sets of equations (i.e. models) that help to provide insight into these problems.

5.3 WHAT WE ARE ATTEMPTING TO ACHIEVE

At this stage it is worth asking the questions: 'What are we trying to achieve in a course on mathematical modelling?'. The most straightforward answer to this question is: 'Our aim in formulating the course from which this book derives was to teach students the basic art of mathematical modelling. Accordingly, our aim for this book is to describe an approach to teaching the basic art of mathematical modelling that has been successful; at least to some limited extent'. By learning the basic art of mathematical modelling we mean that, by the end of the course, students should:

  (i) have the confidence to attempt to make sense of an ill-posed problem with a large amount of mostly irrelevant data;
 (ii) be able to rationalise the ill-posed problem into an unambiguous question to be answered that utilises a subset of the data available;
(iii) be able to develop a mathematical description of the system that will enable a valid answer to be generated to the above question;
 (iv) be able to solve the resulting set of equations and substantiate the validity of the model;
  (v) be able to make predictions using the model which answer the question posed in (ii) above;
 (vi) be able to prepare a written report on the modelling exercise including recommendations, and in addition make a verbal presentation defending the model and its predictions.

As is clear from this list, the putative modeller must be able to do rather more than merely formulate and solve a set of equations.

5.4 A PHILOSOPHICAL APPROACH

Essentially, our approach comes down to providing a working environment which is similar in form to that of the practising modeller. This is because intuition plays a substantial role in mode building and novice modellers must learn to use their intuition in life-like circumstances if their learning experience is to be of any lasting benefit. What does this mean with regard to the classroon situation? For a start it means that the problems as prescribed are not clear, i.e., they are ill-posed and have a lot of irrelevant data, as in most real-life modelling situations. It also means that the students work from within their own range of experience and expertise as do practising modellers. In other words, we do not expect freshmen students to generate finite element solutions to a problem involving stress anslysis. However, they should be able to integrate most straightforward functions.

Important features in the environment of all practising modellers are the available computing facilities and associated numercal software. Up until very recently this would have been a problem for novice modellers with little or no computing background. However, with the advent of interactive 'user-friendly' mini- and micro-computers this factor need no longer be a constraint. In fact, with a modest amount of effort, some automatic equation solvers can be made available as interactive packages that require a minimal amount of computer experience. In the course of our work we have found two types of packages of particular value. For many problems the mathematical model either comes down to a set of ordinary differential equations or an algorithm involving a discrete set of activities. Whilst students can soon develop the ability to formulate reasonably straightforward sets of ordinary differential equations, they usually have difficulty in solving all but the most trivial ones. However, they will quickly grow to appreciate an automatic solver which generates numerical solutions for any prescribed set of parameters and initial values. Provided that the user interface to the solver is friendly enough, students will soon pick up the confidence to generate their own solutions. The same is true for discrete models, Provided that a user-friendly package is available, students will soon grow in confidence enough to use it, both to implement their discrete algorithm and to generate computer results. This utilisation of computing resources in generating model solutions is important because it reflects one of the main reasons why modelling is such a popular activity today. Utilising software packages does not make modelling a soft option; on the contrary, it is a legitimate tool of all modellers. Using the computer 'for real' contributes to the breadth of expertise acquired by would-be modellers.

There are now a growing number of useful computer packages on the market for mini- and microsystems. However, in Chapter 7 we present a summary discussion of the approach that we have taken to the provision of suitable computing facilities and software. With regard to the process of modelling *per se,* there will be no recourse to systems theory although, of course, the modelling approach summarised in Chapter 4 is systematic. The difference between our systematic approach and that of formal systems theory is distinct and meant to be so. We have attempted to develop a pragmatic, informal means of describing the activity of modelling rather than a formal highly structured approach to the subject. The reason for this is simple — we do not believe that any significant work on the modelling of a process or system has been achieved merely by formalising the modelling approach. Good models are produced when the effort is placed on gaining a proper understanding of the process or system under consideration. That is, understanding the problem is kept pre-eminent in the modeller's mind.

Finally, modelling practitioners usually build models from a suitable background of relevant mathematical expertise. That is, they are not usually wrestling with basic mathematical concepts at the same time as trying to obtain a conceptual understanding of the system they are intending to model. For this reason we do not believe that modelling is a suitable medium through which to teach mathematics. It is our experience that such approaches tend to confuse students, simply because they have too many concepts to grapple with at once. This is not to say that learning mathematics and the skills of modelling are conflicting. Indeed, each helps the student to have a better understanding of the other. After all, the student who uses a differential equation 'in anger' as part of a model will certainly learn more about the behaviour of differential equations as an incidental to the modelling exercise. Conversely, a deeper understanding of mathematical concepts can help to give the modeller a much improved appreciation of his system or process. In short, we feel that the activity of mathematical modelling merits attention from mathematicians, engineers and scientists without the encumbrance of having to learn the relevant mathematics simultaneously.

5.5 A TEACHING PROGRAMME

Since intuition cannot be taught, but has to be acquired by relevant experience, it must be encouraged by the framework of any course on mathematical modelling. Furthermore, since the objectives in a modelling course are rather different from other mathematically related disciplines, the student's learning environment must reflect this, that is, it should be one that motivates individuals to participate actively in the modelling process.

Essentially the course has three main parts. The first comprises an introduction to explain what mathematical modelling is and its growing influence in society, and a description of the process of mathematical modelling. Since the ability to model competently is closely related to experience there can be no substitute for it in a course. In the second part of the course, therefore, experience is forced onto the students by presenting them with ill-posed problems and setting them off to determine a solution. Since it is both unfair, unproductive and unrealistic to expect students to think out solutions in a vacuum, they are provided with something like the advantages afforded to the professional modeller, that is, they are set to work in groups so that they can converge to an understanding by discussion. It should be recognised that this is not an instantaneous process. Initially the discussion tends to be very slow. However, as the group begins to sort out ideas and concepts, the discussion becomes rather more audible and animated! The group work is useful because it helps to engender basic confidence in each student by utilising a blend of abilities. Often the more mature students are happy with sorting out concepts, whilst the younger ones will enjoy manipulating the resulting mathematical equations. The third part of the course is categorised by two factors:

(i) the use of computer packages to implement models and generate solutions;
(ii) whilst still discussing the original problem in groups students proceed to model development and subsequent stages individually.

The latter factor ensures that students cannot rely on others to execute some of the aspects of modelling that they do not like or cannot do.

Finally, it should be noted that the work on each model divides naturally into three main parts:

(i) presentation of the problem;

(ii)   model development by the group or individual involving relatively little interaction with the teacher;
(iii) a debriefing session by the teacher consisting of an assessment of the students' approach plus the presentation of his own model.

The following scheme is based upon a twenty-four-week course at 2 hours per week which are taken at one sitting. It assumes that the students have been introduced to the computer (interactive mini-computers to be precise) and that they have become familiar with some ideas of computer programming (quite possibly in BASIC – a language despised by most computer scientists yet very useful as a means for teaching at junior levels).

The course is divided into two terms, possibly taken as two separate courses in succeeding years of a degree.

Term 1

Week 1:   An overview of mathematical modelling, i. e. what it is, how it has developed and why, what it is now used for and why it is so important for our society today? The reasons why student mathematicians, scientists and engineers should be exposed to modelling are outlined (see Chapter 1).

The class is divided into groups of about six to discuss in what sectors of society they are aware of mathematical models in use, as acquired from general information sources (vis, television, popular science magazines, etc.). After some discussion each group reports their findings.

For homework the students find some published reference to the use of mathematical modelling and prepare a short essay on the application and use of the published model.

Week 2:   A description of the process of mathematical modelling as discussed in Chapter 4 is given. The four basic stages of modelling are emphasised; in particular, the importance of the gestation stage is highlighted as the means by which we reduce the initial problem to the one that is to be solved, i. e. the discarding of the irrelevant data, the evolution of a conceptual framework and the formulation of an unambiguous question to be answered. The unambiguous question to be answered leads naturally to an objective to be satisfied. This objective forms the focal point for the model building phase. It is then stressed that the modelling exercise is greatly assisted by continually bearing in mind the phrase, 'What am I trying to do?'.

Finally, students are warned that mathematical modelling is hard work, there are no magic formulae and there is no guaranteed route to success. They should expect to feel unsettled, uncomfortable, confused and lost at various stages of any project. This is part of the learning experience associated with mathematical modelling that experienced modellers suffer from too! However, sticking at the problem, and really *thinking* about it usually yields fruit.

At the end of this discussion students are again divided into groups. The *Apple problem* described in section 6.2 is presented and the groups commissioned to work on it. Homework involves preliminary work on model development.

Week 3: Groups complete model development, generate solutions and arrive at conclusions. Each group's approach to the problem is then presented by one of its members. The homework involves the production of one report per group which is handed in a couple of days before the next class.

Week 4: Initially the teacher debriefs the class on the completed project. Included in the topics discussed are the approaches developed by the various groups, comments on their progress to the point where they decide what the problem really is (i.e. the unambiguous question to be answered, the objective to be satisfied), the actual model, its predictions, their conclusions and recommendations, their presentation and report. A general reflection on the way each group tackled the problem is helpful. This includes highlighting how well each group proceeded together plus a reinforcement of what each group needs to concentrate on to improve their skills. The teacher's own model (e.g. section 6.2(c)) is presented as a (possible!) alternative approach. Although the teacher's presentation standard should be high, his approach should not be presented as *the* solution by any means.

The Iceberg problem (section 6.3) is then presented by the teacher and the groups reconvene to discuss the problem and, one hopes, to reach a suitable stage so that various modelling approaches may be considered as homework.

Week 5: Initially, the groups report back to the class on what they feel the question to be answered is and what objective they have selected to satisfy. Brief discussion then ensues on each group's report to help ensure that all the arising questions and objectives are reasonable.

The rest of the session involves model building in groups whilst the homework probably consists of attempts to solve the model equations.

Week 6: Each group discusses their array of models and comes up with an agreed group model. The model results are then generated and some conclusions reached on the basis of these results.

The class closes with a session where a member from each group presents the work completed by the group. The homework involves writing up a group report on the model development, results and consequent recommendations, etc. Again the report is handed in a couple of days before the next class.

Week 7: Initially the teacher debriefs the class on the Iceberg problem, discusses the models and performance of each group, much as in week 4, and gives a presentation of his own model.

At this stage, it is useful to generate a class discussion on how the students feel about the activity of mathematical modelling. This is a good point at which to have such a discussion because the students will be reasonably familiar with the process of modelling and now have some experience of trying to model. This is a time to elucidate the class's problems and to encourage the students with the fact that experienced modellers also have much the same difficulties that they themselves are now experiencing.

Week 8: Presentation by the teacher of the problem, 'The true cost of a mortgage' (see section 6.4). The class is divided into groups, not larger than three, where at least one member has some programming experience. For the last part of the class a discussion is instigated on what the class feels the problem really is? The homework involves model formulation.

Week 9: Initially, each group meets to discuss their formulations and to agree on one. The rest of the class is spent in developing a solution procedure. For homework the group generates and assesses their model's predictions.

Week 10: Groups meet to discuss and agree on an interpretation of model results. Having done this the class meets to hear a verbal presentation from each group. The rest of the class and homework involves the preparation of a report on the problem by each individual student. Again the report is handed in a couple of days before the next class.

Week 11: Initially the teacher debriefs the class on their performance on the Mortgage model. Here it is useful to have a couple of the best students present their models; during this time it is just as important to expose the motivation behind the model building process as it is to obtain an elegant model. The teacher also outlines his model and gives a copy to the students.

The term is concluded with a teacher's review of how he assesses the progress of the class. Depending on their background and experience, some students are much happier with discussing ideas and developing a sound conceptual view of a situation whilst others feel more comfortable manipulating equations. Each is encouraged to develop his weaker areas.

Term 2

Week 1: The first part of the term concentrates upon a model which uses ordinary differential equations. To ensure that students can actually formulate simple ordinary differential equations, it is useful to formulate and solve in the steady state, a simple model which describes the concentration of salt in a tank of water which retains constant volume and has salty water of variable concentration flowing in.

The Pollutants problem (see section 6.5) is then presented and the class divided into groups of three to discuss it. The session ends with a class discussion on deciding what the problem really is and what question the model will be built to answer. The homework involves formulating the model on an individual basis.

Week 2: The formulation of the model is completed on an individual basis. At this stage the student should have a set of equations that he does not know how to solve. This is, therefore, a convenient point to spend some time introducing a suitable computer package for automatically solving *simple* sets of ordinary differential equations.

The package we have used is called IPSODE and it is described in Chapter 7. After a simplified discussion of the numerical solution of ordinary differential equations and a short account of IPSODE the students are introduced to the package at the terminal. The homework consists of using IPSODE to generate suitable sets of results.

Week 3: This class involves further use of IPSODE to generate and then assess model predictions. It is useful to spend part of the time having a class discussion on what students are looking for in generating numerical results and what can be said about the system and the utility of both dynamic and steady-state solutions to the differential equations.

Week 4: Complete the assessment of model results and arrive at a recommendation based upon them. The preparation of a report is commenced which is completed and handed in before the next class.

Week 5: Three or four students are selected to present their work on the pollution problem and their recommended solution. The problem is discussed with the class and any difficulties highlighted, especially any associated with using the computer. Then the teacher presents his model and distributes the relevant report.

Week 6: An informal approach to discrete simulation is introduced which is oriented more towards communicating the basic ideas rather than delivering a formal lecture. A simple queueing problem is discussed and solved in a straightforward but laborious manner. A suitably easy-to-use computer package is then introduced which allows the user to define the system and automatically generates solutions. We have used APHIDS in our work and this is described in Chapter 7. The students are introduced to the package at the terminal and shown how to solve the simple problem considered above.

Week 7: The Computer Scheduling problem described in section 6.6 is presented and the class divided into groups to discuss the background to the problem. After each group has reported back on their findings and conclusions about the nature of the problem, the students work on the formulation of the model individually. The formulation is completed during the following week.

Week 8: The simulation package **APHIDS** is utilised to implement the model and generate some computer results. A suitable 'complete' set of results may require further work after the class.

Week 9: At this stage much of the assessment of the computed results takes place and conclusions are reached. The report preparation then begins and is completed as homework.

Week 10: Three or four students are selected to present their models and to discuss their results and conclusions. The teacher then debriefs the class and discusses their performance on the project and outlines the development, results and conclusions from his model.

Finally, the course should finish with a class discussion to highlight how the students think the course has gone – its weaknesses, what they found easy or hard, etc., and what could be done to improve the course.

5.6 TEACHER'S ROLE AND ASSESSMENT

From the scheme outlined above there are two factors which have to be considered. It is clear that our concept of a course on mathematical modelling is not for the most part amenable to the talk-and-chalk type of approach. So what is the teacher's role here and how does he interact with the class? The short answer to this question is: as little as possible during the modelling phase of any problem. In practice, the teacher presents the problem outline and ensures that the groups or individuals actually get to work. He chairs the class discussions and report-back sessions and obviously has to carefully prepare his own report on how each group performed on the development and use of any particular model. In general, however, the teacher should keep as low a profile as possible because students are always liable to take anything a teacher says as *the truth*. Since the object of the exercise is to provide an environment where the students are forced to think and act from their own initiative, it is counterproductive to give 'easy' clues – which negates the whole process. A general guideline for the teacher when teaching modelling could be summarised as, 'If in doubt do not say or do anything'. Of course, there are some occasions when students have to be helped back in the 'right' direction, but through experience we can say they are relatively few and far between. In the same way the 'teacher's' model is not meant to be taken as *the truth* with regard to the model and its inclusions; however, the report and level of presentation should give a lead with regard to the standards expected.

The reader may well have noticed that at the beginning of section 5.5 it was stated that the course was supposed to be 24 weeks long and yet only 21 weeks are timetabled. This was done deliberately; the teacher should not worry about appearing to 'process' the number of models through too slowly. It is much more important for the students to go through the whole experience of modelling a few times rather than be helped through a large number of models that may look more impressive on a syllabus! It is from the painful activity of building models themselves that the students most benefit and that experience should not be denied them. However, if the class does make good progress then it is worth setting an individual modelling assignment to each student and giving them a few weeks to complete it. The classes then essentially become a discussion forum for the student to clarify his ideas with his colleagues and teacher. Obviously, the teacher must be careful to ensure that he is acting more as an intelligent library than as the senior and, therefore, dominant partner in the model building exercise.

Assessing the individual student's performance is not an easy matter especially during the first part of the course. Here the only hard information consists of the short essay written at the beginning of the course, the group-generated reports plus the individual report at the end of the term. Because, this information is relatively sparse it is vital that the teacher spends as much time observing each group from a distance to discern who is and, perhaps more importantly, who is not making any real contribution. As an initial attempt it is suggested that each project report is marked according to the following scheme:

Aspect	%
Problem understanding	30
Model formulation	25
Model analysis	20
Report and presentation	25

In addition, the teacher should make an assessment of whether each student is hardworking, average or largely non-participative. Thus,

for the group projects an individual score may be computed from the
following weights:

Student classification	Weight
Hardworking	1
Average	0.8
Non-participative	0.5

Assuming the pass mark to be 40% over the course, this means that
even though two students are nominally responsible for producing
the same report one can achieve a 75% score whilst the other is
deemed unsatisfactory. Obviously, this scheme relies more heavily
on the subjective judgement of the teacher than more conventional
courses. An alternative means of assessing student participation is to
get each group either to distribute, say, 100 marks amongst themselves
or to award their own weights between 0 and 1 to each member.
Experience of colleagues elsewhere indicates that this method works
quite effectively.

When this course was initially operated at Sunderland the complete assessment for the first part of the course was a combination
of course marks and a question or two within a formal examination.
However, the latter is not really very satisfactory and more thought
must be given to achieving a reliable assessment of the student's
progress and a test of his ability to develop mathematical models.

By comparison, there are few problems with making a reasonably
reliable assessment of the student's progress and achievements in the
second part of the course. Here much of the work is carried out
on an individual basis and so the reports should represent a reliable
reflection of what the student actually achieved. Furthermore,
because of the open-ended nature of the problem specification, over-
zealous collaboration or copying can be tested for quite reliably. The
student who really built and analysed a model will have a sound
understanding of what his concept of the problem is all about. In
general, experience confirms that copiers soon show their relative
lack of understanding upon questioning by the teacher. Since the
written work of each student is a reasonably reliable reflection of
his work then continual assessment is probably the most suitable
means of examining a class for the second part of the mathematical
modelling course.

What is clear from the above set of comments on both the teacher's role and assessment is that he must be prepared to move from the dominant role in the classroom to a relatively passive one, spending more time observing the class at work than addressing them. Such a role is entirely in keeping with our philosophical approach to the teaching of mathematical modelling outlined above, i.e. create a realistic environment and the means to enable the students to learn the art of mathematical modelling by experience.

5.7 A SALUTARY EXPERIENCE REVISITED

Now that we have described an approach to the teaching of mathematical modelling that we feel has worked successfully, this is a useful point to compare it with the experience described in Chapter 3. Starting with the positive features: the teacher spent relatively little time, at least initially, on formal presentation. Instead he tried to communicate the basic ideas of modelling, emphasising the factors involved in the process of abstraction from reality. Similarly, the Potato problem was also a good introduction to modelling as it provides some basic experience for the students. However, the failure of the teacher to initially instil a systematic approach into the class and then to reinforce it during a debriefing session obviously means that this first modelling experience was not as beneficial as it might have been. By 'systematic' we here mean getting a good conceptual view of the problem at hand, an unambiguous question to be answered and a criterion to be satisfied, since it is these factors that lead naturally into model formulation, etc.

Another positive feature was the willingness of the teacher to respond to the student's demand for more reality. Unfortunately, the first major mistake here was to overdo it by setting a problem that was too hard for the class involved. After all, the students had very little experience of mathematical modelling and no real idea of a systematic approach. These factors combined with their poor mathematics background, meant that they were ill-prepared for such a large jump in problem complexity and so they were lost from the very beginning. This is unfortunate because it is imperative that enough confidence is generated within the students early on and then constantly reinforced so that they will be able to cope with an increasingly complex set of unfamiliar circumstances which comprise the sequence of modelling problems, the present course attempts to engender and reinforce both the confidence and the useful experience upon which to further develop modelling skills. In short, the complexity jump from the Potato problem to the Supermarket problem was too large and too early in the course.

Even so, there were some other factors that could have improved the class's performance. The teacher invoked a class discussion on the supermarket problem and after this he divided them into groups and excused them a week's attendance to gather data and work together. It probably would have been better to divide the class into groups first and invoke group discussions followed by a report back session. This should have involved more people in trying to understand what the problem means. Furthermore, they would have had some time to become acquainted as a group so that they might work more effectively together outside class. By comparison the groups who took the course described earlier in this chapter often met at least once a week outside the class. The students should not have been excused the next class, but it should have met and after some initial group discussion, been required to make a short presentation on their understanding so far.

After four weeks the active students in the class handed in reports which were classified as reasonable attempts. The teacher then described his own approach to the problem. Although it is obviously a good idea to consider one service point in isolation to model the flow/queue problems through each till, the teacher's desire to get amongst some mathematics rather overtook the need to see the full problem in perspective. Of course, he tried to do this at a later stage, but was too late as, by then, students were generally disenchanted and the classes had reverted to chalk-and-talk lectures. The disenchantment felt by the students was further accelerated by the teacher's inappropriate use of complex mathematics for the type of class involved. It would have been much better to concentrate from the beginning on the factors that affect the profitability of a supermarket, then the flow problems through a cash till would have arisen naturally. In other words, the evaluation of an unambiguous question to answer and a criterion to be satisfied would have naturally steered the students in the right direction.

Finally, moving back to the positive side, the students' complaints at the end of the course really serve to demonstrate that the problem as presented exhibited all the main features of those often asked in real life.

We can summarise what the main aims of the teacher should be as:

(i)   instill a systematic approach of problem formulation into the students at the beginning of the course;

(ii) set problems which engender and reinforce the confidence of students so that they will be motivated to tackle unfamiliar problems of increasing degrees of complexity;

(iii) structure the group work and class discussions in a way that encourages individual participation and effective group working;

(iv) do not try to use mathematical techniques beyond the level of the students involved; and, finally,

(v)   if the classes degenerate in a series of chalk-and-talk lectures then the students are not acquiring the skills of mathematical modelling.

Obviously the approach to teaching mathematical modelling outlined in this chapter attempts to avoid these pitfalls. However, the success of any course obviously depends upon the teacher and how he approaches the task. There is no getting away from it, the approach described here represents a great deal of hard work for both the teacher and the students. A course based upon this approach is certainly no easy option for either teacher or student, but can be a rewarding and valuable experience for both if performed correctly.

CHAPTER 6

The mathematical models

6.1 BACKGROUND NOTES

The purpose of this chapter is to present the modelling scenarios as they should be used. The response of the students to the problem as presented and how they work their way through is also of interest and so a short section on this is included. In many ways this is a useful guide for the teacher so that he has some idea of what to expect when he 'sets his class away' on a problem. A suitable mathematical model for each problem is described, along with some relevant predictions and recommendations. The models described in this chapter are meant to provide an example set of material suitable for use in the type of modelling course proposed in Chapter 5. Other new publications which provide a suitable source of problems are summarised in section 8.2.

6.2 BUYING APPLES

6.2.1 Introduction

TINNO is a company manufacturing a variety of tinned fruits. The low cost of their tinned apples makes it one of their best selling lines. Although the buyer tries to ensure that the quality of apple used is fairly consistent his overriding concern is to minimise purchasing costs. The buyer usually selects from one of two grades – the larger are characterised by an average size of 6 cm diameter whilst the smaller are about 4 cm across. There are about four large apples per pound, which currently cost 24 p/lb, whereas the smaller apples cost 22 p/lb.

Obviously, the factory manager prefers the larger size apples since both the relative amount of waste is less (i.e. 10% by volume for the core and a 2 mm thick peel), and the overall processing time is shorter. However, the buyer must purchase the apples which are most economical — so which are the most economical?

6.2.2 The response

The initial discussion by the group was rather slow; however, most of the groups became rather more animated when they were reminded of the time limit on their discussion time. As the discussions advanced the students argued about what factors were and were not relevant to the matter at hand. These factors included matters such as apple quality differences in the grades A and B. But the most popular debate was on whether to include the cost of any extra work in processing the grade B apples. If this is to be included, then how? After agreeing on what the important factors were, the groups then began to discuss what they thought the problem really amounted to.

In the report-back session a member from each group summarised their discussion pointing out the factors they thought were relevant plus their view of what the problem was. By this stage some groups had specified an unambiguous question to be answered, an objective to be satisfied, and had even made a list of suitable assumptions, whilst others were at much earlier stages of the modelling process. As a result, some of the slower groups benefited from those who had made greater progress. However, the slower groups could not progress much further without really understanding the problem at hand. This was reflected in the following class where the models were due to be completed. Perhaps one of the slow groups still has not come to grips with the problem — so now is the time when they have to work. For most of the class, however, they are now confidently working towards the completion of their model and the generation of a suitable set of results.

In the final report-back session this confidence is reflected in the presentation — they have obtained an answer and feel strong enough to support it. Often their presentations are very informal and, initially, rather disorganised; this is something worth picking up early on.

During the course of the model building exercise an observer should notice a gradual growing in confidence of the group as they master the problem in hand and begin to generate a solution.

6.2.3 An acceptable model

Buying the most economic apples in the sense viewed by the purchaser comes down to determining which grade of apple yields the largest usable volume of apple per unit cost. He is not concerned with the processing costs *per se* and so it does not enter his calculations. In order to develop a model to evaluate this, it is necessary to make some assumptions to make the problem practicable:

(i)   the apples are all spherical,
(ii) they all have a similar quality with no blemishes,
(iii) the core amounts to 10% volume of any apple,
(iv) the peel is 0.2 cm thick everywhere,
(v)   the cost of processing is the same for the two apple grades.

For each grade we have to evaluate the usable volume of apples per pound, since we already know the cost of each grade of apples/lb. Thus, since for each apple there are two sources of waste (i.e. the core and the peel), then net apple volume = gross apple volume − core volume − peel volume. Now the core volume amounts to 10% of the gross volume whilst the peel volume is well approximated by $4\pi r^2 d$ where r is the apple radius and d is the peel thickness. Hence, the net volume of usable apple in an apple is given by

$$V_{net} = \frac{4}{3}\pi r^3 - 0.1.\pi \frac{4}{3}r^3 - 4\pi r^2 d$$

$$= 4\pi r^2 (0.3r - 0.2) \text{ cm}^3$$

where $d = 0.2$ cm.

So if there are n apples/lb, then the volume of usable apple/lb is given by

$$V_u = 4\pi n r^2 (0.3r - 0.2) \text{ cm}^3/\text{lb}$$

Furthermore if C is the cost of the apples/lb then the volume of useable apple per unit cost is

$$V_e = 4\pi n r^2 \{0.3r - 0.2\}/C_r \text{ cm}^3/\text{p}$$

For the grade A apples $r = 6$, $n = 4$, $C = 24$ so that

$$V_e(A) = 120.6 \text{ cm}^3/\text{p}.$$

For the grade B apples we know that $r = 4$ and $C = 22$. However we do not yet know the value of n, i.e. the number of grade ▮ apples/lb. What we do know is that the grade A and B apples have similar quality so that it is reasonable to assume they have the same density, ρ. Therefore, we know that the number of grade B apples/lb is given by,

$$4 \cdot \frac{4}{3} \pi r_{A}^{3} \rho = n \frac{4}{3} \pi r_{B}^{4} \rho = 1$$

i.e. $n = 4 \left(\frac{r_{A}}{r_{B}} \right)^{3} = 13.5$

since $r_{A} = 6$ and $r_{B} = 4$ cm.

Hence, the volume of usable material per unit cost for the grade ▮ apples is given by

$$V_{e}(B) = 123.4 \ \text{cm}^{3}/\text{p}.$$

The results from this model clearly demonstrate that according to the criteria used by the buyer the smaller grade B apples are the most economical. However, it is worth commenting that the difference in usable volume between the two grades is less than 2%. These calcula tions were made assuming that the cost of processing is not affected by the grade of apple. Surely this cannot be true. It must take more effort to decore 13+ small apples as compared to 4 large apples no matter what their size. So a more practical recommendation migh be to set a minimum difference between $V_{e}(A)$ and $V_{e}(B)$, the calcu lated usable volume per cost for each grade, which ensures that the overall benefits of using the smaller apples are really worthwhile.

Note. The observant may have noted that, even with the asumption made above, the calculations are not quite correct! Modellers make implicit assumptions all the time in building models. One hopes they have little effect on the overall conclusions which may be drawn from a model. The reader may be interested to make a more accurate calculation of the total waste, based upon the same assumptions used above, and confirm the results of the author's model.

6.3 ECONOMICS OF MOVING ICEBERGS

6.3.1 Introduction

The cost of desalinating sea water conventionally in the Persian Gulf is high (viz. 10 p/m^{3}) and requires extensive amounts of oil. Some ime ago scientists suggested that it could well prove both practically feasible and less expensive to tow icebergs from the Antarctic, a distance of about 9600 km. Although some of the ice would undoubtedly melt it was thought that a significant enough proportion of the iceberg should remain intact right to the Persian Gulf.

A programme of work was then carried out to evaluate the practical problems associated with such a proposal and to quantify he factors that were likely to influence the economics of such a venture. Amongst other factors was identified the variability in the rental costs of the different-sized towing vessels plus the maximum loads they are able to tow (see Table 6.3.1). It was also found that the melting rate of the iceberg depends upon the towing speed and its distance from the South Pole, at least, up to 4000 km away. Table 6.3.2 summarises the data available to assess the rates at which icebergs melt. Finally, the fuel cost was found to be heavily dependent on the towing speed and the size of the iceberg, though it was relatively independent of the size of the towing vessel. The available data relating to fuel costs is summarised in Table 6.3.3.

Bearing in mind that 1 m^3 of ice only produces 0.85 m^3 of water, evaluate a strategy to produce the cheapest water for the Persian Gulf by towing icebergs and decide on its economic feasibility.

Table 6.3.1   Towing vessel data

Ship size	Small	Medium	Large
Daily rental (£)	4.00	6.20	8.00
Max. load (m³)	500,000	1,000,000	10,000,000

Table 6.3.2   Melting rates of icebergs (m/day)

Towing speed km/h \ Distance from Antarctic (km)	0	1000	>4000
1	0	0.1	0.3
3	0	0.15	0.45
5	0	0.2	0.6

Table 6.3.3 Fuel costs (£/km)

Towing speed km/h	Iceberg vol. (m³) 10^7	10^6	10^5
1	12.6	10.5	8.4
3	16.2	13.5	10.8
5	19.8	16.5	13.2

6.3.2 Student response

This problem was rather more demanding than the first, a fact which was reflected in the initial discussion. The groups were happy to quickly get to the root of the problem and sort out an unambiguous question to be answered and an objective or criterion to be satisfied. However, even with a week or so to mull things over, they found it hard to go beyond general mathematical statements in the model development phase. This was because the students knew that relationships existed amongst the various factors that influence the overall transport costs, yet they were not obvious from the tabular information that was available. At this stage, some time was taken to explain how empirical equations may be fitted to the data. The derivation of one of the equations was given in detail and the students were left to compute the other. For the anxious teacher, this does not make the problem much easier, but merely facilitates the construction of a suitable model. Having acquired the necessary relationships the groups then proceed with their model development. When a group has settled upon a model it is useful to spend a few minutes with them to help ensure that it is consistent.

The final report-back session should be more organised than that associated with the first problem and the students should have a better idea how to structure their presentation.

6.3.3 An acceptable model

The problem as specified requires us to determine the towing conditions which minimise the cost of producing water in the Persian Gulf via icebergs and to decide whether it is cheaper than conventional production methods. Essentially, this comes down to evaluating the cost/m^3 of usable water for any specified set of towing conditions. Therefore we have to evaluate both the final iceberg volume and the total transportation cost.

To even begin building a model of this system a number of simplifying assumptions have to be made, viz.:

  (i) icebergs are spherical,
  (ii) the melt rate is the same at all points of the iceberg,
  (iii) icebergs move at uniform speed,
  (iv) weather conditions do not influence the problem,
  (v) the towing vessel operates ideally (e.g. no strikes, breakdowns, etc.),
  (vi) raw fuel costs remain constant during the journey,
  (vii) changes take place at slow enough rates to be able to assume constant conditions for each day.

Clearly, these idealisations will affect the model results. However, it is hoped they will not affect the form and order of magnitude of the model's predictions. Having said this, of course, it does mean that the model predictions will have to be treated with some caution.

It is clear from the comments above that to achieve our objective we must:

  (i) evaluate how much of the original volume arrives intact;
  (ii) estimate the total cost to transport the arrival volume of water.

Naturally, the volume of ice lost on the journey depends on the iceberg melting rate. However, from the supplied data the melting rate is affected by both the towing speed and the distance from the Antarctic. From Table 6.3.2 some thought leads to the conclusion that for a distance less than 4000 km from the Antarctic the melting rate may be described by an equation of the form

$$R_m = (a + bl)(1 + cu) \qquad \text{m/day} \qquad (6.3.1)$$

Substitution of the data in equation (6.3.1) leads to the following parameter values: $a = 0$, $b = 5.10^{-5}$, $c = 0.2$. Beyond a distance of 4000 km from the Antarctic the melt rate is independent of l and so

$$R_m = \begin{cases} 5.10^{-5}l\,(1 + 0.2u) & <4000 \text{ km} \\ 0.2\,(1 + 0.2u) & \text{otherwise} \end{cases} \text{(m/day)} \qquad (6.3.2)$$

Thus, if $R(0)$ is the original radius of the iceberg with initial volume $V(0)$, then after i days, the radius becomes

$$R(i) = R(i-1) - R_m(i) \qquad \text{(m)} \qquad (6.3.3)$$

and the volume reduces to

$$V(i) = \frac{4}{3}\pi R(i)^3 \qquad \text{(m}^3\text{)} \qquad (6.3.4)$$

Furthermore, to calculate $R_m(i)$, we need the distance from the Antarctic,

$$l(i) = 24\,iu \qquad \text{(km)} \qquad (6.3.5)$$

The total number of days to complete the journey,

$$d = \frac{9600}{24u} = \frac{400}{u} \qquad \text{(days)} \qquad (6.3.6)$$

combined with equations $(6.3.2)-(6.3.5)$ constitutes a simple model to evaluate the final volume on arrival in the Persian Gulf.

Turning attention now to the costs of transporting the iceberg, this consists of two parts, the daily rental and the fuel expenses. From the data available it is clear that the fuel cost depends upon both the towing speed and the size of the iceberg. In a similar way to the evaluation of the melt rate, the fuel cost may be estimated from the following formula,

$$c_f = 0.3(u+6)(\log_{10} V - 1) \qquad \text{(£/km)} \qquad (6.3.7)$$

The daily transportation cost may then be estimated as

$$c_d(i) = R + 7.2u\,(u+6)(\log_{10} V(i-1) - 1) \quad \text{(£/day)} \quad (6.3.8)$$

where R is the daily rental of the towing vessel. The total transportation cost is

$$C_T = \sum_{i=1}^{d} c_d(i) \qquad (6.3.9)$$

which leads us to the cost of 1 m³ of usable water as

$$C_W = \frac{C_T}{0.85\ V(d)} \tag{6.3.10}$$

Equation (6.3.10) yields the result we seek, i.e. the cost of producing 1 m³ of water in the Persian Gulf via the transportation of an iceberg. However, equations (6.3.2)−(6.3.10) constitute the set of model equations required to evaluate C for any set of specified conditions. It is possible to solve the model equations using a hand calculator, although it is probably easier to generate results by writing a short computer program. The results shown below were, in fact, generated via a program written in BASIC on a PDP 11/40 minicomputer.

The computer program evaluated the final iceberg volume, total transportation cost and, for each size vessel, towing speeds from 1 to 5 km/h and a series of allowable initial iceberg volumes. The computer results for the largest vessel are summarised in Table 6.3.4. In the case of the small and medium towing vessels, the concomitant small initial iceberg sizes give rise to relatively quick melting. Therefore the iceberg either melts before the journey is over or so little arrives that the resulting water is relatively expensive to produce. In fact, only the largest towing vessel appears to be economically feasible. Even in this case, however, the model predicts that volumes in excess of 7 million m³ must be towed at speeds greater than 3 km/h if the exercise is to be feasible.

In assessing the utility of the model predictions it is worth bearing in mind the assumptions inherent in developing the model. For example, although the assumption of a spherical shape greatly simplified the mathematics it also overestimates the volume of ice that will arrive intact. Considering these and other assumptions it is probably not worth considering iceberg towing as a practicality unless there are a set of operating conditions for which the cost is much less than the 10 p/m³ of conventional methods. Since the predicted minimum cost is around 5.5 p/m³, then the viability of the proposal must be considered marginal at best. If, however, the relative cost of water rises sharply with respect to that of oil then the proposal may become more obviously economic.

Table 6.3.4 Model results for largest vessel

Initial volume ($m^3 \times 10^5$)	Speed (kw/h)	Final volume (m^3)	Cost total (£)	Cost/volume (£/m^3)
10	1	797618	413288	.609592
10	2	$.297069.10^7$	283679	.112345
10	3	$.416622.10^7$	251121	.070912
10	4	$.484509.10^7$	244306	.049322
10	5	$.529404.10^7$	246695	.054822
9	1	620746	412000	.780843
9	2	$.253318.10^7$	282438	.131171
9	3	$.361498.10^7$	249790	.081292
9	4	$.423407.10^7$	242853	.067479
9	5	$.4645 \ .10^7$	245115	.062082
8	1	461817	410531	1.04582
8	2	$.211244.10^7$	281040	.156518
8	3	$.30781 \ .10^7$	248294	.094900
8	4	$.36359 \ .10^7$	241222	.078052
8	5	$.400775.10^7$	243343	.071433
7	1	322671	408825	1.49059
7	2	$.171092.10^7$	279441	.19215
7	3	$.25578 \ .10^7$	246587	.113419
7	4	$.305258.10^7$	239363	.092251
7	5	$.338416.10^7$	241325	.083894
6	1	205459	406795	2.32933
6	2	$.133181.10^7$	277574	.245198
6	3	$.2057 \ .10^7$	244603	.139897
6	4	$.248681.10^7$	237204	.112218
6	5	$.277674.10^7$	238984	.101255
5	1	112630	404291	4.223
5	2	979378	275334	.330743
5	3	$.157967.10^7$	242233	.180404
5	4	$.194223.10^7$	234632	.142124
5	5	$.218891.10^7$	236197	.126949
4	1	46793.7	401035	10.0827
4	2	659645	272542	.486075
4	3	$.113155.10^7$	239297	.248796
4	4	$.142419.10^7$	231453	.191194
4	5	$.162566.10^7$	232757	.168444
3	1	9966.28	396389	46.7918
3	2	381632	268843	.828773
3	3	721507	235445	.383911
3	4	941033	227297	.284165
3	5	$.109487.10^7$	228270	.245282
2	1	35.525	387949	12847.6
2	2	160124	263397	1.93525
2	3	364879	229866	.741149
2	4	507439	221314	.513104
2	5	610548	221830	.427446
1	1	*** Iceberg melted after 342 days		
1	2	22689.3	253131	13.1252
1	3	93983.6	219772	2.75107
1	4	155457	210634	1.59404
1	5	203739	210415	1.21502

6.4 ESTIMATING THE TRUE COST OF A MORTGAGE

6.4.1 Introduction

Imagine the following scenario some time in the mid-1980's. The manager's office of a well-known building society contains three people, Mr and Mrs Smith and the manager himself, Mr Loansome. As we join them the following (typical) conversation is taking place.

MR SMITH: We would like to buy a house, but don't really know much about the availability of mortgages or how much we can borrow. Can you help us?

MR LOANSOME: Since you have saved with us for some time, Mr Smith, there should be no trouble in arranging a mortgage for you. The questions you have to consider are what size mortgage you need and which type of mortgage you require.

MR SMITH: We'd obviously like to borrow as much as possible and I'm afraid we know very little about the different types of mortgage.

MR LOANSOME: Let's deal with the amount you wish to borrow first. We normally allow married customers to borrow about two and a half times their joint income. Can you tell me your incomes?

MR SMITH: We have a small child so my wife doesn't work and I earn around £12,000 a year as an engineer.

MR LOANSOME: So that means you can borrow about £30,000 and since most mortgages are 80% of the house price, you should be looking for a house at around £37,500.

Let us now turn to the type of mortgage. Basically, there are two kinds of mortgage. In the conventional mortgage you borrow a sum at an annual interest rate (e.g. 10% at present) over a specified term, usually 25 years. We then evaluate a monthly payment in which you pay off a combination of interest and capital. Obviously, at the beginning your are paying mostly interest, whilst at the end it's mostly capital. You also have to take out a basic life insurance policy to cover the mortgage in the event or your death which costs about £4/year/£1000 borrowed. Thus, on a £30,000 loan your gross monthly payment comes to around £300. However, to help encourage buyers, the government gives full tax relief (e.g. 30%) on the mortgage interest payments, while insurance premiums are net of tax relief. In the early stages this reduces your net monthly payment to about £200. In addition, for loans of £30,000 or less the borrower has the tax relief deducted at source through the MIRAS scheme. In this way, the borrower does not receive tax relief on his income, but makes a reduced monthly payment of equivalent value. In this way the repayment scheme is structured so that the net payment is constant over the loan period.

Turning now to the second kind of mortgage – the endowment – here you still borrow a sum of money over a specified term, but at a slightly higher interest rate (e.g. 10.25% at present). However, you don't pay back the capital directly. Instead, you pay the interest on the loan and take out an insurance policy which matures either at the end of the specified term or on the death of either of you. Depending on the policy you can balance the endowment so that the policy benefits either just pays off the loan *or* makes a certain level of profit. For example, a policy costing £5/year/£1000 borrowed yields zero profit; if the payment goes up to £15/year/£1000 borrowed the profit is £300 per thousand borrowed and if the payment is £20/year/£1000 borrowed then the profit is £1000 per thousand borrowed.

Your monthly payment, therefore, has two components, consisting of the interest on the borrowed capital plus the net of tax cost of the insurance policy. This makes your gross monthly payment around £315 whilst the tax relief brings this down to about £210.

As you can see, there are two main noticeable differences between the mortgage schemes.

(a) the gross cost of the conventional mortgage is smaller;
(b) at the end of the term the endowment mortgage leaves you a nice nest-egg.

Our friends may now respond as follows.

MR SMITH: My first reaction is that the monthly payment seems quite high for both schemes though I can now see why it must be, for us to earn 10.5% interest on our savings.

MR LOANSOME: Yes, the cost appears high, but you must remember your income is likely to increase by at least the annual rate of inflation whilst the payments stay the same so that after a few years the cost becomes relatively small.

MR SMITH: Yes, I suppose so. But which of these mortgage schemes is the best?

Yes — which is the best?

6.4.2 Student response

At this stage the class should be divided into groups and dispatched to discuss the relative merits of the mortgage schemes. In particular, they should be asked to decide upon a criterion for 'best'.

After some group discussion (for half-an-hour or so) the lecturer should then ask each group to summarise the results of their discussions. This will probably involve three main aspects — establishing an understanding of the way the mortgage schemes work, identifying the factors that influence their cost and the selection of a suitable choice criterion.

Most will soon establish all the possible factors that can influence mortgage costs (including acts of God) and will also suggest ignoring many of the more nebulous ones (e.g. moving house, non-payment, etc.) in order to reduce the difficulty of the problem. During the last decade inflation has featured prominently in most people's financial dealings and they will almost certainly wish to retain this effect.

Deciding upon the criteria for 'best' is not straightforward as there are a number of factors to consider. Based upon past experience students have come up with a number of criteria, for example,

(a) investing the difference of the cheaper mortgage each year and evaluating the returns at the end of the term;
(b) evaluating the ratio of overall return : total cost;
(c) calculating a measure of the real cost by summing the yearly fractional costs of each mortgage, as devalued by inflation.

The students generally find that coming to a decision over a suitable criterion is difficult. They also find it conceptually difficult to sort out the effects of inflation from the rest of the problem. It is sometimes useful, after the students have agonised for a while (i.e. an hour or so), to suggest that first they tackle the problem without inflation and introduce its effects later. The students then proceed with their selection of a suitable criterion for best and proceed to develop their model.

6.4.3 An acceptable model

In the following, one particular way of developing a model is described. It may well not be the best, but appears adequate in some non-quantifiable sense.

In order to simplify matters it seems reasonable to assume

(i)   all the interest, insurance and tax rates are constant,
(ii) all payments are made on time,
(iii) no house resale,
(iv) no disasters (e.g. fire).

The criterion to be used in evaluating the best mortgage is (c) above — that is, calculating the real cost by summing the yearly fractional cost of each mortgage (with respect to gross income) and adding any financial returns at the end of the term to the last year's income.

(a)   Endowment mortgage model

The annual payment for the endowment mortgage has a number of contributory factors, viz.

[annual net cost, P] = [interest costs] − [tax relief] + [insurance policy (net of tax)]

Hence,

$$P = MI_e - TMI_e + SM$$

$$= M[I_e(1 - T) + S]$$

where the symbols are explained in Table 6.4.1.

If the fractional annual inflation rate is r, then the cost of the mortgage in real terms in year i is given by

$$C(i) = \frac{P}{(1 + r)^{i-1}}$$

The actual cost of the mortgage will be

$$T_n = nP - BM$$

where BM represents the lump sum received at the end of the term. The devalued cost of the mortgage is given by

$$T_d = \sum_{i=1}^{n} P/(1 + r)^{i-1} - \frac{BM}{(1 + r)^{n-1}}$$

$$= \frac{P}{r(r + 1)^{n-1}}[(1 + r)^n - 1] - \frac{BM}{(1 + r)^{n-1}}$$

Table 6.4.1 Nomenclature

Item	
B	– fractional profit on endowment investment at term
C_i	– amount of capital paid off in year i
$C(i)$	– annual cost of endowment in real terms
I	– annual fractional interest on repayment mortgage
I_e	– annual fractional interest on endowment mortgage
M	– total sum borrowed
n	– period of loan
$P_i, \bar{P}$	– annual net payment on repayment mortgage
$\bar{P}(i)$	– annual cost of repayment mortgage in real terms
P	– annual net payment on endowment mortgage
r	– annual fractional rate of inflation
S	– annual endowment policy costs (£/£ borrowed)
S_r	– annual insurance policy costs on repayment mortgage (£/£ borrowed)
T	– annual rate of taxation
T_n	– actual total cost of mortgage
T_d	– total cost of mortgage with allowance for inflation

(b) Repayment mortgage model

In this case, the factors that contribute to the net cost of the mortgage are summarised as follows:

[annual net cost in year i, P_i] = [mortgage payment] + [insurance cost (net of tax)] − [tax relief on mortgage interest]

An additional constraint is that the net payment is constant over the whole period. Hence, we may write for year 1,

$$P_1 = C_1 + MI(1 - T) + S_r M$$
$$= C_1 + MV$$

where $V = I(1 - T) + S_r$ and C_1 is the amount paid off the capital loan. In year 2

$$P_2 = C_2 + (M - C_1)V$$

and for year i

$$P_i = C_i + (M - \sum_{q=1}^{i-1} C_q)V$$

Since, the repayment is constant over the borrowing period then,

$$P_1 = P_2 = \ldots = P_i = \bar{P}, \text{ a constant}$$

Hence, for example,

$$C_1 + MV = C_2 + (M - C_1)V$$

i.e.

$$C_2 = C_1(1 + V)$$

and

$$C_i = C_1(1 + V)^{i-1}$$

By noting that $\displaystyle\sum_{i=1}^{n} C_i = M$ it may be shown that

and that

$$C_1 = VM/[(1 + V)^n - 1]$$

$$\bar{P} = \frac{MV(1 + V)^n}{[(1 + V)^n - 1]}$$

Now, the annual payment when devalued by inflation becomes

$$\bar{P}(i) = \frac{\bar{P}}{(1 + r)^{i-1}}$$

Thus, the actual total cost of the mortgage may be given by

$$T = n\bar{P}$$

whereas, the cost when devalued by inflation becomes

$$T_d = \bar{P} \sum_{i=1}^{n} 1/(1 + r)^{i-1} = \frac{\bar{P}}{r(r + 1)^{n-1}}[(1 + r)^n - 1]$$

If it is assumed Mr Smith borrows £30,000 for 25 years, where all the above data is used as relevant, the following table of results may be generated if 5% inflation is assumed:

Scheme	T (without bonus)	T (actual)	T (devalued)
Repayment	66694	66694	39478
Endowment-NP	57562.5	57562.5	34074
Endowment-LP	65062.5	65062.5	35722
Endowment-FP	68812.5	38812.5	31430

The results are interesting in that before any profits are paid at the end of the loan period, it is apparent that the no-profits scheme is the cheapest in terms of pound notes actually paid out. However, once the bonus is accounted for, whether or not inflation is accounted for, it appears that the full-profits scheme is quite the best. The repayment and low-profit endowment are not as good as the other two schemes in any respect. Hence, on the basis of the model above the following recommendations are made.

(i) If the smallest outlay at the time is necessary then the no-profits endowment scheme is best.
(ii) If the client can afford the larger repayments then the lump sum at the end of the loan period makes the full-profits endowment policy the best financial value.

6.5 MINIMISING WATER POLLUTION

6.5.1 Introduction

The new steelworks now under construction at Throup, Surrey, wants to discharge polluted water into the River Semidee. The local council stipulate that this can only be done provided an effluent system is installed which ensures that the pollutant concentration entering the river is less than 5.10^{-4} g/m^3 under all circumstances. This demand presents a problem to the technical manager, Mr T. I. Mill, because he knows the plant discharge concentration is likely to vary between 10^{-3} and 10^{-2} g/m^3. Although the local council put a fair amount of pressure on companies to install equipment which comprehensively treats effluent under all operating conditions, they tend not to be so strict on companies once the plant is working. They generally issue warnings or small fines for offences and only close plants down temporarily if they exceed the legal limits excessively.

Mr Mill's R&D department have advised that biological treatment systems are likely to be the most reliable in dealing with the kind of pollutant the steelworks will produce.

At this stage Mr Mill approached a number of companies for estimates of possible systems. Poogone Inc. came back with a design for a single well-mixed tank system. However, their recommended volume was 10,000 m^3 and this, together with the ancillary equipment, would require more space than is currently available. Mr Mill felt that the polluted water must be able to be cleaned much more efficiently as the flow rate into the system would only be 10 m^3/h. He therefore asked his R&D department two basic questions:

(i) Will the proposed design always work?
(ii) Could a more reliable system using two much smaller tanks in series be designed?

Apart from the flow dimensions, there was some basic data available on the growth and digestive characteristics of the biological organisms in the treatment tank. It turns out that the organism death rate is 10^{-5} h whilst the growth rate of the organism population is directly proportional to the pollutant concentration (with a proportionality constant of 1.2625 m^3/h). Finally, in a similar way the pollutant consumption rate is directly proportional to the organism concentration with a proportionality constant of 0.1 m^3/g h.

Imagine you are part of the R&D team given the above task. Build a model to help Mr Mill.

6.5.2 Student response

This problem has relatively little in the way of irrelevant information. However, this is more than made up for by the requirement to formulate a set of differential equations to describe one- and two-tank problems. Although the students will have had some experience of formulating differential equations they will still find this problem less than trivial. They will also have to spend some time in deciding exactly what analysis has to be done on the model equations to yield the required answers. However, they soon grow in confidence when a steady-state analysis of the single tank case shows that the Poogone Inc. design will not work if the incoming effluent density reaches the upper end of its range. Formulation of the two-tank case is not generally so difficult although the steady-state analysis is normally thought-provoking for most students. Finally, performing the dynamic analysis using the interactive computer package IPSODE also proves to be an interesting and rewarding experience for the students.

6.5.3 An acceptable model

Polluted water is discharged from the steelworks at Throup and the management have been ordered by the authorities to install an effluent system to 'clean up' the pollution before the water enters the River Semidee. In particular, this system must ensure that the concentration of pollutant is less than 5.10^{-4} g/m^3 at all times. Now this could be a problem since the plant discharge concentration varies between 10^{-3} and 10^{-2} g/m^3.

Biological treatment systems are generally considered to be the most reliable at dealing with the steelworks' pollutant and one company (Poogone Inc.) has submitted a design utilising a single well-mixed tank with a volume of 10,000 m^3 coping with a flow rate through the system of 10 m^3/h. Some concern was expressed by the steelworks management at the tank size recommended by Poogone. This concern generated two questions:

(i) Will the proposed scheme always work?
(ii) Could a more reliable system using two much smaller tanks be designed?

The purpose of this report is to provide some answers to these questions, and to make a recommendation on the system to be used and its dimensions.

The single-tank system is the most straightforward and is shown in Fig. 6.5.1. It consists of a well-mixed tank containing a volume V of water which has uniform concentrations of pollutant, $c(t)$, and organisms, $B(t)$. Polluted water, with a concentration c^* flows into the tank at the rate Q m^3/h and clean water flows out at the same rate.

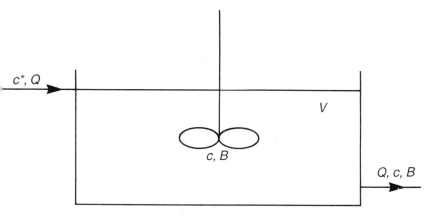

Fig. 6.5.1 – The single-tank system.

Some basic microbiological investigations have revealed the following data on the growth and digestive characteristics of the organisms. The rate at which organisms multiply is directly proportional to the pollutant concentration with a proportionality constant of $R_2 = 1.2625$ m^3/h. The associated death rate constant is given by $D = 10^{-5}$/h. The rate of pollutant digestion by the organisms is assumed proportional to the organism concentration, and the relevant proportionality constant is $R_1 = 0.1$ m^3/g h.

A pair of ordinary differential equations may be constructed to describe the variations in pollutant and organism concentrations.

For the pollutant, in the time interval $(t, t + \Delta t)$

$$\begin{array}{ccc} \text{change in pollutant} \\ \text{level in tank} \end{array} = \begin{array}{c} \text{pollutant arriving} \\ \text{from inflow} \end{array} - \begin{array}{c} \text{pollutant leaving} \\ \text{in outflow} \end{array}$$

$$- \begin{array}{c} \text{pollutant digested} \\ \text{by organisms} \end{array}$$

which may be written mathematically as

$$V(c(t + \Delta t) - c(t)) = Q c^* \Delta t - Q c(t) \Delta t - R_1 c(t) B(t) V \Delta t$$

i.e. $\dfrac{dc}{dt} = \dfrac{Q}{V}(c^* - c) - R_1 B c$ (6.5.1)

For the organisms, their corresponding change is defined by:

$$\begin{array}{ccc} \text{change in organism} \\ \text{population in tank} \end{array} = \begin{array}{c} \text{organisms created} \\ \text{from reproduction} \end{array} - \begin{array}{c} \text{organisms lost} \\ \text{via death} \end{array}$$

$$- \begin{array}{c} \text{organisms washed} \\ \text{out in the outflow} \end{array}$$

which may be written mathematically as

$$V(B(t + \Delta t) - B(t)) = R_2 B(t)c(t)V\Delta t - DB(t)V\Delta t - QB(t)\Delta t$$

i.e. $\dfrac{dB}{dt} = B(R_2 c - D - \dfrac{Q}{V})$ (6.5.2)

Equations (6.5.1) and (6.5.2) describe the way in which the organism and pollutant concentrations vary in the tank. Now although there is a great deal of interest in devising efficient ways to start-up, etc., initally at least the basic question may be taken to mean that once a desirable operating condition has been reached will the proposed volume (i.e. 10,000 m^3) be adequate? If it is, how well will the system cope with changes in the incoming pollutant concentrations, c^*? The first of these questions may be answered by looking at the steady-state solutions of equation (6.5.1) and (6.5.2) (i.e. when $\dfrac{dc}{dt} = \dfrac{dB}{dt} = 0$). These equations give two solutions

(a) $c = c^*$ and $B = 0$

(b) $c = (D + \dfrac{Q}{V})/R_2$ and $B = \dfrac{Q}{V}(c^* - c)/cR_1$

The first solution obviously corresponds to a 'washout' situation and is undesirable! Using the available data, the second solution implies

$$c = \left(10^{-5} + \frac{10}{10^4}\right)/1.1625$$

This means that the basic size of the tank is too small and will never reduce the pollutant to below the specified concentration of 5.10^{-4}. The tank volume required to do this at steady-state would be

$$V = Q/(R_2 c - D) = \frac{10}{(1.2625 \times 5.10^{-4} - 10^{-5})} = 16{,}096 \text{ m}^3$$

This volume assumes no variation in the incoming pollutant concentration, c^*. The most dramatic variations in c^* could be either from 0.01 to 0.001 or vice versa. The single tank system was simulated using IPSODE and the tank's behaviour for the change from 0.001 to 0.01 is illustrated in Figs 6.5.2 and 6.5.3. From this result it is

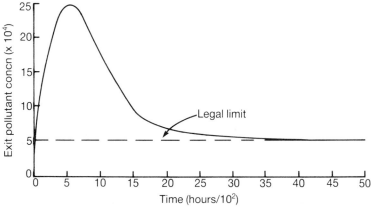

Fig. 6.5.2 – Response of single-tank with volume 16,130 m³ to discharge pollutant level change 0.001 → 0.01.

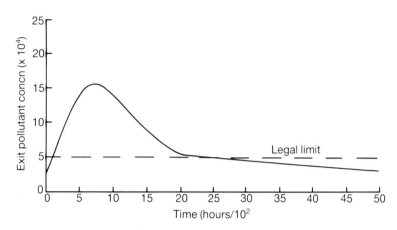

Fig. 6.5.3 – Response of single-tank with volume 30,000 m³ to discharge pollutant level change 0.001 → 0.01.

clear that although eventually the pollutant level falls below the legal limit there is a substantial time period where c is greater than 5.10^{-4} (i.e. up to 10,000 hours). In theory, a larger tank should help to damp out such excursions. Various sizes were tried. However, even when the volume, $V = 30,000$ m³, the pollutant level still exceeds the legal limit, for 2,000 hours, as shown in Fig. 6.5.3. Although, this is an improvement it still fails to meet the legal criteria. Furthermore, since the tank size necessary to avoid exceeding the legal limit (i.e. $c < 5.10^{-4}$) at all times is now very large and will exceed the originally estimated size by an order of 5 (at least), the single tank system is considered as unsatisfactory.

The two-tank system is illustrated in Fig. 6.5.4. The tanks are assumed to be of volume V_1 and V_2 and the outflow from tank one constitutes the inflow of tank two. Furthermore, there is no reason

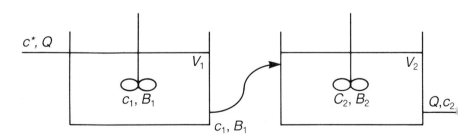

Fig. 6.5.4 – The two-tank system.

to suppose that the growth and digestive behaviour of the organisms is any different from that for the single-tank systems.

The behaviour of the organism and pollutant concentrations in each of the tanks may be described by a system of four ordinary differential equations. The mathematical description of the first tank is formally the same as for the single-tank system. Hence,

$$\frac{dc_1}{dt} = \frac{Q}{V}(c^* - c_1) - R_1 B_1 c_1 \tag{6.5.3}$$

$$\frac{dB_1}{dt} = B_1(R_2 c_1 - D - \frac{Q}{V_1}) \tag{6.5.4}$$

The change in pollutant concentration in tank two is given by

change in pollutant level in tank two	=	pollutant arriving from inflow	−	pollutant leaving in outflow pollutant
		−	pollutant digested by organisms	

i.e

$$\frac{dc_2}{dt} = \frac{Q}{V}(c_1 - c_2) - R_1 B_2 c_2 \tag{6.5.5}$$

The corresponding variation in the organism population is given by

change in organism population in tank two	=	organisms created from reproduction	−	organisms lost by death
+	organisms arriving inflow from tank one	−	organisms lost in the outflow	

i.e.

$$\frac{dB_2}{dt} = B_2(R_2 c_2 - D) + \frac{Q}{V_2}(B_1 - B_2) \tag{6.5.6}$$

Equations (6.5.3), (6.5,4), 6.5.5) and (6.5.6) constitute the model for the two-tank system. The problem now comes down to using these equations to evaluate V_1 and V_2 in such a way that $c < 5.10^{-4}$ under all possible variations in operating conditions.

As for the single tank, it is useful to look at the steady-state situation for some basic design criteria. Equations (6.5.3) and 6.5.4) yield the same solutions for B_1 and c_1 as in the single-tank case, i.e.

(a) $c_1 = c^*$ $B_1 = 0$

$$\tag{6.5.7}$$

(b) $c_1 = (D + Q/V_1)/R_2$ $B_1 = \frac{Q}{V_1}(c^* - c_1)/c_1 R_1$

Equation (6.5.7) places a constraint on the minimum size of tank one.

Obviously $c < c^*$, hence (6.5.7) implies

$$c^* > (D + \frac{Q}{V})/R_2$$

i.e. $V_1 > Q_1/(R_2 c^* - D)$ (6.5.8)

If $c^* = 0.01$ then $V_1 > 792.1$ m whilst if $c^* = 0.001$ then $V_1 > 7984$ m
If V_1 is less than either of these values (for corresponding stable conditions) then $c_1 = c^*$ and $B = 0$, and the pollutant passes through untreated. It is useful to consider why this is so. From equation (6.5.4) it is clear that if $\frac{Q}{V_1}$ is too large it will dominate over the organism population growth rate (i.e. $R_2 c_1$) and all the organism will be washed out in the outflow. It is this effect which gives rise to the minimum viable volume for the first tank. Assuming, the pollutant is treated in the first tank, the solution of the second pair of equations yields

$$B_2 = \frac{Q}{V_2} \frac{(c_1 - c_2)}{R_1 c_2}$$ (6.5.9)

If it is assumed that $c_2 = 5.10^{-4}$ (i.e. the prescribed upper legal limit on the pollutant level discharged to the river), then equation (6.5.9) (with only V_2 unknown) may be substituted into the steady-state form of equation (6.5.6), i.e.

$$\frac{(c_1 - c_2)}{R_1 c_2} (R_2 c_2 - D) + (B_1 - \frac{Q}{V_2} \frac{(c_1 - c_2)}{R_1 c_2} = 0$$

This equation may be manipulated to give

$$V_2 = \frac{Q(c_1 - c_2)}{(c_1 - c_2)(R_2 c_2 - D) + B_1 R_1 c_1}$$ (6.5.10)

where, for a given value of V_1, B_1 and c_1 may be calculated from equation (6.5.7) and all the other parameters are known.

A small program was written to calculate V_2 for a range of V values for steady-state conditions at the extremes (i.e. when $c^* = 0.01$ and 0.001). A listing of the program and the results for the two cases are shown in Fig. 6.5.5. From these results it is clear

```
LIST
10   REM PROGRRAM TO CALCULATE THE VOLUME OF THE SECOND TANK
20   PRINT "      VOLONE CONONE BUGONE BUGTWO VOLTWO"
30   F$="   #####.# .##### .##### .##### #####.#"
100  D=1E-5: R2=1.2625 : R1=0.1 : C2=5E-4
110  Q=10
120  C9=0.01
130  FOR I=1 TO 10
140  V1=I*200+600
150  C1=(D+Q/V1)/R2
160  IF C1>C9 THEN C1=C9
170  B1=Q*(C9-C1)/(V1*R1*C1)
175  V2=Q*(C1-C2)/((C1-C2)*(R2*C2-D)+B1*R1*C2)
180  B2=(Q/V2)*(C1-C2)/(R1*C2)
190  PRINT USING F$,V1,C1,B1,B2,V2
200  NEXT I
210  STOP
220  END


RUN
     VOLONE CONONE BUGONE BUGTWO VOLTWO
      800.0 .00991 .00115 .11805 15939.9
     1000.0 .00793 .02612 .11843 12545.8
     1200.0 .00661 .04277 .11866 10295.6
     1400.0 .00567 .05464 .11883  8694.3
     1600.0 .00496 .06355 .11894  7496.7
     1800.0 .00441 .07047 .11903  6567.1
     2000.0 .00397 .07600 .11909  5824.6
     2200.0 .00361 .08052 .11914  5217.9
     2400.0 .00331 .08428 .11917  4712.9
     2600.0 .00305 .08746 .11920  4285.9

STOP  @      210

120  C9=0.001
130  FOR I=8 TO 16
140  V1=I*1000

RUN
     VOLONE CONONE BUGONE BUGTWO VOLTWO
     8000.0 .00100 .00002 .00621 16032.3
     9000.0 .00089 .00140 .00622 12471.6
    10000.0 .00080 .00250 .00623  9634.7
    11000.0 .00073 .00340 .00623  7319.7
    12000.0 .00067 .00414 .00623  5393.5
    13000.0 .00062 .00477 .00623  3764.6
    14000.0 .00057 .00531 .00622  2368.2
    15000.0 .00054 .00577 .00622  1156.9
    16000.0 .00050 .00618 .00621    95.6

STOP  @      210
```

Fig. 6.5.5 – Listing of DESIGN program plus the results on calculation of V_2 for the incoming pollutant level at 0.01 and 0.001.

that the value of c^* has a critical effect on the tank sizes required to treat the pollutant satisfactorily. Although a value of $c^* = 0.01$ yields much smaller combined tank volumes than for the single volume system, it may not be practicable. Taking as an example $V_1 = 2,600$, the total volume $< 7,000$ m^3. However, if c^* changed suddenly to

0.001 then a slow washout would take place from both tanks since each of their volumes is less than the viable volume of 7984 m corresponding to $c^* = 0.001$. This means that all designs must be based upon a c^* value of 0.001 so that the worst cases can always be dealt with. This obviously means using the second set of results shown in Fig. 6.5.5. From these results it is clear that in any two-tank system, the steady-state analysis would indicate that a large first and small second tank are desirable. The question thus remains – how does the two-tank system respond to a step change in c^* from 0.001 to 0.01? This change is again used since it is the most dramatic that can occur.

To answer the above question it was proposed to evaluate the steady-state values of c_1, B_1, c_2 and B_2 when $c^* = 0.001$, and then to run the simulation using these steady-state values as the initial values and to use $c^* = 0.01$. This would then simulate a process which had been in a steady-state situation for a while with $c^* = 0.001$, suddenly having a step change in c^* to 0.01.

If

$$\frac{dc_1}{dt} = \frac{dB_1}{dt} = \frac{dc_2}{dt} = \frac{dB_2}{dt} = 0$$

this should provide four equations from which the steady-state values of c_1, B_1, c_2 and B_2 may be evaluated.

Equation (6.5.7) yields the useful steady-state values for c_1 and B_1.

Equations (6.5.5) and (6.5.6) then yield

$$\frac{Q}{V_2}(c_1 - c_2) - R_1 B_2 c_2 = 0 \qquad (6.5.11)$$

$$B_2(R_2 c_2 - D - \frac{Q}{V_2}) + B_1 \frac{Q}{V_2} = 0 \qquad (6.5.12)$$

Everything is known in these equations save B_2 and c_2. B_2 may be eliminated by expressing it in terms of c from equation (6.5.11) (i.e. $B = Q(c_1 - c_2)/R_1 c_2 V_1)$ and substituting into equation (6.5.12),

$$\frac{Q}{V_2}\frac{(c_1 - c_2)}{R_1 c_2}(R_2 c_2 - D - \frac{Q}{V_2}) + B_1 \frac{Q}{V_2} = 0$$

i.e.

$$(c_1 - c_2)(R_2 c_2 - D - \frac{Q}{V_2}) + B_1 R_1 c_2 = 0 \qquad (6.5.13)$$

This equation is a quadratic in c_2 and may be rearranged to give

$$R_2 c_2^2 - (c_1 R_2 + D + \frac{Q}{V_2} + B_1 R_1) c_2 + c_1 (D + \frac{Q}{V_2}) = 0 \quad (6.5.14)$$

Hence, in the general case,

$$c_2 = \frac{1}{2R_1} \left\{ (c_1 R_2 + D + \frac{Q}{V_2} + B_1 R_1) \pm \right.$$
$$\left. [(c_1 R_2 + D + \frac{Q}{V_2} + B_1 R_1)^2 - 4c_1 R_1 (D + \frac{Q}{V_2})]^{1/2} \right\} \quad (6.5.15)$$

```
LIST
10 REM PROGRAM TO GENERATE INITIAL CONDITIONS FOR IPSODE
20 D=1E-5 :R2=1.2625 :R1=.1 :C2=5E-4
30 Q=10
40 C9=.001
15 PRINT " VOLONE    VOLTWO   CONONE   BUGONE   CONTWO   BUGTWO   CONTRE   BUGTRE"
50 FOR I=1 TO 11
60 READ V1,V2
70 C1=(D+Q/V1)/R2
80 IF C1>C9 THEN C1=C9
90 B1=(C9-C1)*(Q/V1)/(R1*C1)
110 B9=R2*C1+D+Q/V2+B1*R1
120 B8=C1*(D+Q/V2)
125 IF B9*B9<4*B8 THEN 163
130 C2=B9/2+0.5*(B9*B9-4*B8)^0.5
135 IF C2>C1 THEN C2=C1
140 C3=B9/2-0.5*(B9*B9-4*B8)^.5
145 IF C3>C1 THEN C3=C1
150 B2=Q/V2*(C1-C2)/R1/C2
160 B3=Q/V2*(C1-C3)/R1/C3
162 GOTO 180
163 PRINT "ROOTS ARE COMPLEX"
164 GOTO 190
180 F$=" #####.#    #####.#   .##### #.#### .##### #.#### .##### #.####"
185 PRINT USING F$,V1,V2,C1,B1,C2,B2,C3,B3
186 NEXT I
190 STOP
195 DATA 10000,7000,12000,5000,13000,4000,14000,3000,15000,2000
196 DATA 16000,1000,14000,5000,14000,6000,14000,7000,14000,10000
197 DATA 14000,14000
200 END
RUN
```

VOLONE	VOLTWO	CONONE	BUGONE	CONTWO	BUGTWO	CONTRE	BUGTRE
10000.0	7000.0	.00080	.0025	.00080	0.0000	.00053	.0072
12000.0	5000.0	.00067	.0041	.00067	0.0000	.00048	.0077
13000.0	4000.0	.00062	.0048	.00062	0.0000	.00047	.0078
14000.0	3000.0	.00057	.0053	.00057	0.0000	.00046	.0079
15000.0	2000.0	.00054	.0058	.00054	0.0000	.00046	.0079
16000.0	1000.0	.00050	.0062	.00050	0.0000	.00047	.0079
14000.0	5000.0	.00057	.0053	.00057	0.0000	.00040	.0085
14000.0	6000.0	.00057	.0053	.00057	0.0000	.00038	.0087
14000.0	7000.0	.00057	.0053	.00057	0.0000	.00035	.0090
14000.0	10000.0	.00057	.0053	.00057	0.0000	.00029	.0095
14000.0	14000.0	.00057	.0053	.00057	0.0000	.00024	.0100

```
STOP   @   190
```

Fig. 6.5.6 − Listing of DESTWO plus results on calculation of C_2 for a range of tank volume combinations.

Another program, DESTWO, was written to evaluate the steady-state values c_1, B_1, c_2, B_2 for any given values of V_1, V_2 and c^*. A listing of this program is contained in Fig. 6.5.6 along with a sample of the more important results. The most crucial variable is c_2 (i.e. concentration leaving the second tank) and these results indicate that $V_1 = 14,000$ m³ may be the most useful value for the volume of the first tank and $V_2 > 5,000$ m³ looks promising for the second. A number of runs using the simulation program were made using initial values of c_1, B_1, c_2, B_2 evaluated from $V_1 = 14,000$ m³ $c^* = 0.001$ and various volumes for the second tank. Of course although, the initial values were evaluated using $c^* = 0.01$ to assess the response of the system.

The results when $V = 7,000$ m³ are illustrated in Fig. 6.5.7, from which it may be estimated that the exit pollutant concentration exceeds the prescribed level by up to a factor of two for about 500 hours. A tank of size 25,000 m³ is necessary to ensure that the pollutant level *never* exceeds the legal limit (see Fig. 6.5.8).

The design of Poogone Inc. recommending a single tank of volume 10,000 m³ is totally inadequate to process the pollutant concentration discharged from the steelworks. If the discharge pollutant level could be kept constant then a tank size of about 16,000 m³ would be required. Since the pollutant level will vary, to avoid exceeding the legal limit for long time periods a very large tank (probably

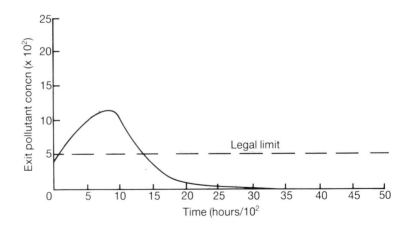

Fig. 6.5.7 – Response of two-tank system ($V_1 = 14,000$, $V_2 = 7,000$) to discharge pollutant level change $0.001 \rightarrow 0.01$.

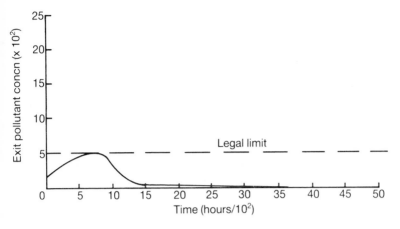

Fig. 6.5.8 – Response of two-tank system ($V_1 = 14,000$, $V_2 = 25,000$) to discharge pollutant level change $0.001 \rightarrow 0.01$.

xceeding 50,000 m^3) would be required. This renders the single-tank ystem completely impracticable.

The equations constituting the model for the two-tank system how that, to avoid washout from tank one, its volume must exceed 3,000 m^3. In fact, the avoidance of washout places a serious cons-raint on the tank sizes that can be utilised. This factor concentrates ttention on the design aspects of the situation when the incoming ollutant level is low (i.e. 0.001), for it is here that the tank volumes eed to be largest. The system then had to be capable of coping vith a step change of 0.001 to 0.01 in the incoming pollutant level. The model showed that it was possible to always keep the exit ollutant concentration less than 5.10^{-4} provided $V_1 = 14,000$ m^3 and $_2 > 25,000$ m^3. However, it was also found that if $V > 7,000$ m^3, hen the exit pollutant concentration was no more than twice the pecified limit for a period of less than 1,000 hours. In practice, it is xtremely unlikely that the steelworks discharge level will move from).001 to 0.01 and stay constant for very long time periods. Assuming hat the discharge pollutant level from the steelworks will vary on week-to-week basis (i.e. every 100 hours or so) it is reasonable to ropose a design consisting of two tanks with $V = 14,000$ m^3 and $= 7,000$ m^3. Although, this is twice the size proposed by Poogone nc. it should be much more effective at processing the pollutant evel.

Finally, it is worth noting that if the steelworks pollutant leve could be maintained near 0.01 then a much smaller system would b￼ required to process it effectively. Therefore it would seem clear tha the criteria used by the designers at Poogone only included $c^* = 0.01$

With regard to recommendations it would seem that one of th￼ following courses of action is the most appropriate:

(i) construct a much larger treatement plant than had originall￼ been envisaged to effectively process the effluent;

(ii) build a smaller plant to make a gesture at effectively treatin￼ the pollutant most of the time and pay the fines when the lega limit is exceeded for long periods;

(iii) investigate the possibility of achieving better control of th￼ pollutant concentration entering the treatment system (i. e￼ nearer 0.01) and, hence, using a much smaller treatment systen altogether.

6.6 AUGMENTING AN INTERACTIVE COMPUTING SYSTEM

6.6.1 Introduction

The scene is Dodge College in the early 1980s. The plannin￼ committee at the college have announced a grant of £20,000 t￼ help improve the performance of the existing computing system Currently the system is considered to be inadequate to meet th￼ college's needs. The committee asked the system manager togethe￼ with a small team to identify how to spend the grant to best effect

The computer, a HYPO-8 operates under a straightforwar￼ interactive system called IBAT. This operating system is really a com bination of fully interactive and batch modes. As such, the systen essentially has two tasks. The first simply involves processing an￼ given job until completion. The second task involves interaction wit￼ users in the provision of editing facilities, calling up files and obey ing instructions to commence the running of a job (which include￼ printing a file). The only devices on the HYPO system are th￼ terminals and the printer. The 'overhead' computing cost consistentl￼ accounts for about 20% of the CPU time available. The termina concentrator has a buffering capability and sends jobs one at a tim￼ to queue for the computer CPU. The system manager decided tha￼ his first problem was to find an effective measure of the HYPO'￼ performance and then to assess the costs and benefits of the availabl￼

selection of possible computer enhancements. At present the college's HYPO-8 consists of a central processing unit (CPU), one storage device, two teminal concentrators (each serving 10 terminals) and two lineprinters. Users log in to the computer at a terminal, manipulate or edit files, run jobs and print their output files onto one of the lineprinters. After a job has been submitted to the computer it joins a queue where it waits for the central processing unit to become available. The CPU only processes one job at a time and then sends the output file to the lineprinter. The choice of lineprinter is based on shortest queue length. The measure of how efficiently a computer operates is normally termed 'performance'. Suitable indicators of performance are usually considered to be a combination of two or more of the following factors: throughput, response time and availability. Throughput means the total number of jobs processed by the computer in a given time period. Response or turn-around means the time lapse between submitting a job to the computer and the results becoming available. The availability of a system is the proportion of nominal working time that it is able to accept jobs. For example, a system with recurrent hardware problems has low availability. According to the system manager's team, who are biased, the HYPO-8 is available 99% of the time and thus availability is not considered as having much influence on performance. After consulting some texts on the subjects the team opted to define a measure of performance as

$$ S = \frac{\text{throughput}}{\sqrt{(\text{average response time})}} \tag{6.6.1} $$

Using this definition when S is large the performance of the computer system is good and when S is small it is poor. In this definition the average response time is a sum of the average waiting time and the average service time.

With regard to computer enhancements it took a month or two and a fair amount of discussion with HYPO's manufacturers to elicit the options available, since they were trying to sell the college a new system! It turned out to be quite feasible to attach a variety of peripheral devices (i.e. terminal concentrations and lineprinters) together with enhancements to the computer memory which will shorten the service times by the CPU. Both a new terminal concentrator and a lineprinter cost out at £10,000 whilst memory enhancements can be implemented at two levels costing £10,000 or £20,000.

Of course, the computer team is interested in comparing the performance of HYPO under a variety of different system configurations. They highlighted the following factors as being the main ones which will influence the computer's performance:

(i) pattern of interarrival times of submission of jobs by users,
(ii) service times of the users,
(iii) the service mechanisms.

Having examined all the available operating data they decided that HYPO users fell into three major categories which they named NOVICE, STUDENT and RESEARCH. Table 6.6.1 summarises the service times required by the terminal concentrator and lineprinters for all three categories. After some consultation with the HYPO manufacturers, the team estimated computer processing times by the CPU for the current configuration and with the possible memory enhancements (see Table 6.6.2). The team then carried out a detailed survey to evaluate the interarrival patterns of all three categories of users. Table 6.6.3 summarises their findings over the whole day when divided into three-hour intervals.

Having acquired all this data one of the team was given the task of producing a model of the various computing configurations possible and then to use it to decide which would best meet the college's future needs. Imagine you are the selected team member, what recommendations would you make regarding the enhancements of the HYPO computer? You should include 'fall back' conditions if any funding difficulties arise.

6.6.2   Student response

If students have not had exposure to a user-oriented interactive discrete simulation then this problem could well appear to be intractable to them. However, assuming that the students have had some experience of using a package like APHIDS, as outlined in Chapter 7, the problem described above should not appear so daunting. In this case, the students soon begin to sort the problem out and crystallise the important aspects with regard to developing a discrete simulation model.

Table 6.6.1 Service times of devices in minutes

	Category of user		
Peripheral device	Novice	Student	Research
Terminal concentrator	0.40	0.75	2.0
Lineprinter	0.80	1.80	5.0

Table 6.6.2 CPU service times (in minutes) for various levels of investment

	Category of user		
Improvement investment (£)	Novice	Student	Research
0	0.32	0.75	3.5
10,000	0.129	0.70	3.3
20,000	0.28	0.68	3.2

Table 6.6.3 Range of user interarrival times (in minutes)

	Category of user		
Time periods in the day	Novice	Student	Research
1900–1200	0.9–1.1	1.8–2.2	13–17
1200–1400	0.7–0.9	0.9–1.1	21–29
1400–1700	0.9–1.1	1.8–2.2	17–23
1700–2100	4.5–5.5	4.5–5.5	9–11

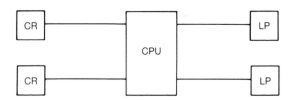

(a) Existing scheme.

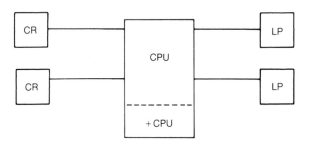

(b) Enhanced CPU at the most expensive level.

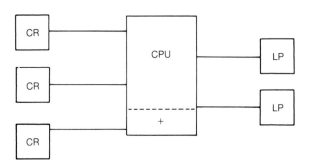

(c) Lower-level CPU enhancement plus terminal concentrator.

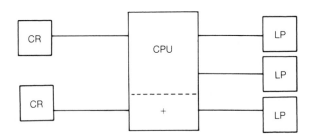

(d) Lower-level CPU enhancement plus lineprinter.

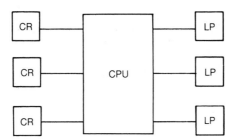

(e) Existing system enhanced by a terminal concentrator and lineprinter.

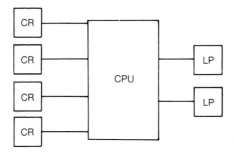

(f) Two new terminal concentrators.

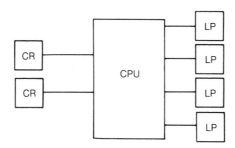

(g) Two new lineprinters.

Fig. 6.6.1 – Existing system configuration and possible enhancements.

6.6.3 An acceptable model

Having sorted through the available information it is clear that the current system has two terminal concentrators, one central processing unit (CPU) and two lineprinters. Figure 6.6.1 illustrates the existing system configuration and the possible enhancements, viz.

  (i) one new terminal concentrator and one new lineprinter,
  (ii) two new terminal concentrators,
 (iii) two new lineprinters,
 (iv) £20,000 worth of memory,
  (v) £10,000 worth of memory and one new terminal concentrator,
 (vi) £10,000 worth of memory and one new lineprinter.

The main objective of the simulation is to effectively compare the performance of the above set of available configurations.

Summarising the available information it appears there are three categories of user whose service times by the terminal concentrators, lineprinters and CPU are summarised in Tables 6.6.1 and 6.6.2 whilst their interarrival times for the whole day are specified in Table 6.6.3. Since data is available for the whole day it seems sensible to use 0900 to 2100 hours as the simulation period. It also appears reasonable to assume the operating period is devoid of breaks since the system's people report that HYPO-8 has 99% availability and the operating overhead is fairly consistent.

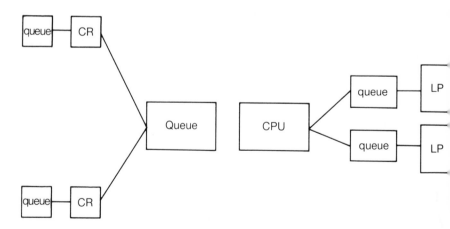

Fig. 6.6.2 — Structure of model.

Using the structure imposed by APHIDS, the flow of traffic is illustrated in Fig. 6.6.2. The system is called the *computer* and it has three categories of server – terminal concentrator, lineprinter and CPU. The customers are called *jobs* and have three categories – novice, student and research. Each *job* can be thought of as having three tasks to complete in order, i.e.

(i)   queue and then be processed by terminal concentrator;
(ii) queue and then be processed by CPU;
(iii) queue and then be processed by the lineprinter.

There are either two or three each of terminal concentrators and lineprinters and jobs join the shortest queue to be served. There is, of course, only one CPU with one associated queue.

Constant interarrival times, as specified in Table 6.6.3, were used and a set of statistics was generated using the APHIDS package under three sets of conditions. These involved using the minimum, maximum and mean values quoted for the interarrival times. It is assumed that the machine stops at 2100 hours regardless of whether or not any jobs are still being processed.

The results of the simulation are summarised in Table 6.6.4. Using the output facilities provided by APHIDS statistics were compiled which accounted for the system throughput (i.e. number of jobs processed), average waiting time by each user before processing commences, the average service time for each job, the total turn-around time for each user plus the measure S of the system's performance. Recalling that a high value for S means good performance, it is clear from Table 6.6.4 that one new lineprinter combined with a £10,000 memory enhancement will be the most effective in improving the computing service, i.e. minimising the turn-around times for the user.

In fact, it may be deduced from the simulation results that the computer's performance should always improve quite dramatically if just one lineprinter were added to the system and nothing else. In other words, the main hold-up in the job production chain is the availability of a lineprinter. The addition of card-readers or CPU enhancements makes no significant difference to the system performance. It is the addition of a lineprinter that reduces turn-around times by a factor of 2 to 3. This is further reduced by up to a factor of 4 when the memory enhancement is included.

Table 6.6.4 System performance under varying conditions

Plan no.	Throughput	Av. wt.	Av. service	Av. response	S
(a) Maximum values of interarrival times					
1	862	5.138	2.618	7.756	309.52
2	862	18.609	2.618	27.227	187.10
3	862	5.009	2.618	7.627	312.13
4	862	18.452	2.554	21.006	188.08
5	862	18.424	2.573	20.997	188.12
6	862	2.748	2.573	5.321	373.69
present	862	18.747	2.618	21.365	186.49
(b) Minimum values of interarrival times					
1	1042	26.481	2.619	29.100	193.16
2	971	36.843	2.608	39.451	154.59
3	1042	26.409	2.621	29.030	193.39
4	968	36.690	2.541	39.231	154.55
5	962	36.738	2.564	39.302	153.45
6	1063	18.075	2.575	20.650	233.92
present	967	36.982	2.616	39.544	153.78
(c) Values of interarrival times half-way between minimum and maximum					
1	951	13.736	2.613	16.349	235.20
2	941	30.528	2.617	33.145	163.45
3	953	15.750	2.630	18.380	222.29
4	941	30.383	2.615	32.998	163.81
5	941	29.617	2.608	32.225	165.77
6	951	7.834	2.613	10.447	294.23
present	941	30.161	2.610	32.771	164.38

On the basis of the simulation work reported in this report the recommendations are simple:

(i) If the full £20,000 is available then a lineprinter and £10,000 worth of memory enhancements will optimise performance;

(ii) if only half the original amount is made available then it would be best invested in an extra lineprinter.

CHAPTER 7

Computer software to support the teaching of mathematical modelling

7.1 INTRODUCTION

It is clear from the rest of this text that computing and computer programming are considered to be an integral part of using mathematics to solve real-world problems. In fact, one aspect of the philosophical approach adopted here involves using the computer as a tool in such a way that it does not materially interfere with the use of mathematics in achieving a satisfactory solution to a problem. For this reason, it is recommended that students use whatever language they feel happiest with. If they are reasonably competent programmers then we would strongly recommend the use of one of the many extended versions of BASIC now on the market. Academic computer scientists may sneer at such suggestions, but it cannot be repeated often enough that in mathematical modelling both the mathematics and any associated computing must be subservient to achieving a solution to the problem at hand. In line with the above philosophy we recommend that any students following a course on mathematical modelling should have an interactive mini- or micro-computer available.

In the rest of this chapter, two computer packages, written in extended BASIC, are described which essentially provide automatic solvers for both continuous and discrete simulation. The packages are user-oriented in such a way as to alleviate the problems associated with generating computer solutions to sets of mathematical equations. The software described in the following pages is available from the authors.

7.2 DEVELOPMENT PHILOSOPHY

Both the simulation packages described in this chapter were developed in extended BASIC on a DEC 11/40 under the RSTS/E operating system and subsequently implemented on a HARRIS 125S under the VULCAN operating system. Both the packages are interactive, permit a high degree of flexibility for the users and enable a full student class to use each package simultaneously. Within the interactive environment the main objectives of the simulation package may be defined as:

(i) easy to access and understand by users with little or no computing experience;

(ii) the package should be foolproof, i.e. fully secure and uncorruptable by user-generated errors;

(iii) user data input errors should be trapped and helpful diagnostics given;

(iv) easy access available to modify data;

(v) provide a comprehensive 'HELP' facility so that the user can obtain information and/or instructions at any stage of the simulation process;

(vi) permit a selection of relevant output information in a variety of ways.

Most of the above aims would apply to any simulation package. In addition, however, these packages are specifically designed to help 'novice' users construct computer models by guiding them through the model building process.

7.3 CONTINUOUS MODEL SIMULATION

A good proportion of mathematical models describing the dynamics of systems really come down to sets of ordinary differential equations. Traditionally any model which consists of a set of ordinary differential equations with time as the independent variable is called continuous simulation. Mathematically, this is expressed as:

$$\frac{dy_i}{dt} = f_i(t, y_1, \ldots, y_n) \quad (i = 1, \ldots, n)$$

where f_i may involve differentials of any or all of the dependent variables. Essentially, the simulation consists of defining a set of initial conditions and then calculating how the dependent variables change with time.

Although this procedure sounds straightforward, there is a large body of engineers and scientists who will testify otherwise. In real systems numerical problems are encountered with ill-conditioned sets of equations and discontinuous changes in system parameters.

Over the years a great deal of pragmatic research has been carried out to develop reliable numerical solution techniques. Currently there is a substantial body of knowledge in this area and the great proportion of continuous simulation problems can now be adequately tackled. Accordingly, a large number of continuous system simulation languages have numerical facilities which enable them to cope with most problems whilst simultaneously masking many of the numerical problems from the user. However, most of these packages are really meant for professionals at mathematical modelling and simulation rather than novices. Here we use the term 'novice' to indicate a lack of mathematical, computational and professional experience. As novices, the students find such packages hard to use and distracting or diverting from the primary objectives in mathematical modelling. Furthermore, it is also unlikely that students will require any of the sophisticated numerical algorithms provided by the packages. However, they would certainly find it diverting if they had to generate numerical solutions to even a small set of ordinary differential equations.

To meet the solution needs of the novice modeller the authors have developed a user-oriented interactive package for solving reasonably well-behaved sets of ordinary differential equations. The package, called IPSODE (Interactive Programming with Systems of Ordinary Differential Equations) is implemented in a multi-user environment and gently steers the novice through the main continuous simulation process, i.e. defining the equations, specifying parameters, values, initial conditions, range of integration, etc.

7.4 IPSODE – INTERACTIVE PROGRAMMING WITH SYSTEMS OF ORDINARY DIFFERENTIAL EQUATIONS

One problem common to all packages is that by the very nature of their generality, they tend to be large thus requiring a large storage space and taking along time to process. The authors tried to avoid this difficulty with IPSODE by subdividing it into two programs,

IPSODE and SECOND. The general description of the problem, i.e. the differential equations themselves, are entered in IPSODE. They are checked as far as possible by the program and then presented for checking to the user and, when verified, they are stored in a program called MODEL in compiler form. A compiled version of SECOND is then attached to MODEL, at which point the student enters actual values for the problem he is solving. Another advantage of this arrangement is that MODEL need only be constructed once for each problem and either many different runs can be made by one student or many students can use it simultaneously.

As the package may cope with any set of non-stiff. linear first-order differential equations, a standardised approach is needed. The user has first to identify dependent variables, independent variables and parameters. For example, if the equations were:

$$A_1 \frac{dm_1}{dt} = Q_1 - Q_2 m_2 \quad \text{and} \quad A_2 \frac{dm_2}{dt} = Q_3 m_1 + m_2$$

then m_1, m_2 are termed dependent variables, t is the independent variable and A_1, A_2, Q_1, Q_2, Q_3 are the parameters. Dependent variables are called Y(1), T is always the independent variable and the parameters are called P(1). Thus, the equations should be entered as:

$$\frac{D(Y(1))}{DT} = (P(3) - P(4) * Y(2))/P(1)$$

$$\frac{D(Y(2))}{DT} = (P(3) * Y(1) + Y(2))/P(2)$$

When this was first tried on students, many of them confused Y(1) and Y(2) so they were asked to give them a name, e.g. MASS1 and MASS2 and the equations then looked like:

$$\frac{D(MASS1)}{DT} = (P(3) - (P(4) * Y(2)))/P(1)$$

$$\frac{D(MASS2)}{DT} = (P(5) * Y(1) + Y(2))/P(2)$$

The left-hand side of these equations are printed by the program, the student has only to enter the right-hand side.

IPSODE permits three types of parameters: constant, varying and dynamically modifiable. Constant parameters are as their name implies. If in the above example Q_1 was a function of t then this is an example of a varying parameter. A parameter may vary (a) continuously, (b) discretely or (c) continuously but with a single discontinuity allowed subject to one condition. Examples of these would be:

(a) $Q_1(t) = 0.3 \sin t$

(b) $Q_1(t) = \begin{cases} 2 & 0 < t < 2 \\ 3 & 2 < t < 4 \\ 4 & 4 < t < 5 \end{cases}$

(c) $Q_1(t) = \begin{cases} -2t & t < 0 \\ +2t + 1 & t > 0 \end{cases}$

Full instructions of how to enter particular cases are given by the program and one of its distinctive features is that by typing a special key, a HELP message is available which gives detailed explanations on any section of the input. A dynamically modifiable parameter is one whose value can be changed whilst the program is still working. If the program was modelling the filling of a bath by a tap and the rate of flow of water is governed by the parameter $P(1)$, then $P(1)$ can be increased or decreased by a fixed amount during the actual solution time; and if the output was in graphical form the effect of turning the tap on or off could be simulated. This is useful for modelling control schemes. An example of IPSODE in use is shown in Fig. 7.1.

SECOND requires the specific values of the parameters, initial values of the dependent variables, the range of integration and the form of program output. An example of SECOND is shown in Fig. 7.2. The first and most simple involves the tabular listing of time (the independent variable) and the dependent variables every ten of the original time steps on the user's terminal. A list of numbers, however, is not always helpful in providing insight for students. Thus the option to plot one up to all of the dependent variables at the user's terminal is provided. In fact the user can opt either to generate the plot dynamically or to plot after a list has been printed and vice versa. The final form of output is to file for storage and possible further processing.

An important question asked in SECOND is 'Do you want to use the automatic stepsize?'. If the student answers YES then the equations will be solved automatically. The solver is in fact a straightforward fourth order Runge–Kutta combined with Richardson's

```
RUN
    If at any stage you wish to change any data already input,press CONTROL+S
    Do you want help with the input of your problem?Y
    For this program it is assumed that the L.H.S. of the
    equations consist of first-order differentials w.r.t. time.
    This is called the INDEPENDENT variable.
    The other variables are DEPENDENT ones.
    If the independent variable is not T but say,X,then you must do a
    simple substitution before using the program.
    The parameters are the numbers and constants in your equations.
    These are to be stored as P(I).
    If your equations contain symbols which have a constant value then
    these values are to be stored as CONSTANT parameters.
    A VARYING parameter is one where the value changes with respect
    to T.It can take the form of a table (discrete type) or it can
    be represented by continuous functions.An example is available.

    How many dependent variables are there  2

    Are there any control variables?N
    Give descriptive names for other dependent variables as Y(I)
    Variable Y( 1 ) is called?CONC
    Variable Y( 2 ) is called?BUGS
    How many equations are there  2

    Except for your declared variables,all other constants and
    coefficients must be input as parameters P(I).
    Do any parameters change their value during the calculation?Y
    How many of these parameters are there  1

    How many constant parameters are there  5

    Please confirm there are 1 varying parameters
    labelled P(1)
    and 5 constant parameters labelled as P( 2 ) to P( 6 )?Y
    Would you like an example of varying parameters?N
    How many varying parameters vary continuously  0

    Please input the equations in the following way
    The L.H.S. is of the form D(Y(I))/DT where I is the number of the equatio
    The R.H.S. is a combination of Y(I),P(I) and T.
    Would you like an example?N
    D(CONC)/DT=?P(2)*(P(1)-Y(1))/P(3)-P(4)*Y(1)*Y(2)
    D(BUGS)/DT=?Y(2)*(P(5)*Y(1)-P(6)-P(2)/P(3))
    Your  2 equations are:-
    1  )  D(CONC)/DT=P(2)*(P(1)-Y(1))/P(3)-P(4)*Y(1)*Y(2)
    2  )  D(BUGS)/DT=Y(2)*(P(5)*Y(1)-P(6)-P(2)/P(3))
    Check these carefully.Are they correct?Y
    Do you wish to change any section?N
    Wait a few minutes while MODEL is being constructed.
    Information has been compressed and put into MODEL
    Type OLD 900LIBA*SECOND then JS MODEL and then RUN
END    @    999
```

Fig. 7.1 – An example of constructing the MODEL program from IPSODE.

```
OLD SECOND
JS MODEL
RUN
    Input of varying parameters.
    Each of these 1 parameters must be given in the form
    of a table;T,V typed on a new line in order of increasing T where
    T is the first time value that the parameter attains the value V
    How many values does P( 1 ) take?2
            T , P( 1 )
    Pair  1 ?200,0.001
    Pair  2 ?20000,0.01

    Have you input the varying parameters correctly?Y
    Input of constant parameters.
    Input the constant parameters as P(I),where I begins at  2
    P( 2 )=?10
    P( 3 )=?16097
    P( 4 )=?0.1
    P( 5 )=?1.2625
    P( 6 )=?0.0001
    Have you input the constant parameters correctly?Y
    The initial value of CONC is?0.001
    Do you wish to restrict CONC?N
    The initial value of BUGS is?0.1
    Do you wish to restrict BUGS?N
Do any of the varying parameters have a non-zero initial value N

    Have you input initial values correctly?Y
    Which interval (A,B) are you considering?
    A= 0

    B= 1000

    Do you want the automatic stepsize?Y
    You are considering the interval ( 0 , 1000 )
    Please confirm?Y
    Give a title to your differential equations problem

BUGSY
    Do you require your output
                1) Stored in a file
                2) As a print-out
                3) In graphical form
    Please type 1,2 or 3   2

    Are there any varying parameters you wish to observe  N
```

Fig. 7.2 – An example of using the program SECOND of the IPSODE package.

extrapolation technique. If the student is numerically competent, he can answer NO and then control the stepsize manually.

The program is fully interactive insofar as it will spot obviously wrong answers to questions and not abort, i.e. it will spot numerical answers to questions expecting a YES/NO and vice versa. It is written so that a general user is guided through the input but there is a facility for getting more help by pressing the CONTROL + C keys at the terminal. This is usually the abort signal in the RSTS-E system but the signal is trapped and instead the user is transferred to the HELP file where he can get help on any section or go back to any section to change input that he has now realised is wrong or inappropriate (an example is shown in Fig. 7.3). The program is watertight insofar as whatever the student types will not result in him being 'thrown out of the system'. There is nothing more frustrating for a

```
There is a facility for choosing a certain parameter and observing
its effect on the graph by changing its value
Do you wish to use this facility?  Y
Which parameter do you wish to vary?  I
You may increase it or decrease it by a fixed amount of?  0.01
You must give the parameter in the form P(I) or P(I%)
Which parameter do you wish to vary?  I
You may increase it or decrease it by a fixed amount of?  0.01
Please supply numerical data
You may increase it or decrease it by a fixed amount of?  0.01
You must give the parameter in the form P(I) or P(I%)
Which parameter do you wish to vary?  CC
Type Y for help : type N for change?  Y
 I.  Control parameters        2.  Varying parameters
 3.  Constant parameters       4.  Initial values of variables
 5.  Range , stepsize          6.  Option of file, print, graph
 7.  Status report             8.  Plot of print-out results
 9.  Repeat solution          10.  Graph of other (or all) variables
II.  Finish off
Which section do you want help on?  Type a number between I and II?  7
A summary of the current parameter values and variables is given
STATUS REPORT FOR BUGGY

Current value of CONC     = .001
Current value of BUGS     = .I
Current value of P(I)     = .01
Current value of P(2)     = 10
Current value of P(3)     = 16097
Current value of P(4)     = .I
Current value of P(5)     = 1.2625
Current value of P(6)     = .0001
```

Fig. 7.3 – Using the HELP facilities of IPSODE.

student than to spend an hour typing in values and then, because of one wrong input value, being aborted by the program and thus losing all his work. Figure 7.3 shows an example of how a student did not understand a question and repeatedly typed in the wrong response. Eventually he typed CONTROL + C and resolved the situation. (This question has since been rephrased to make its meaning clearer.)

One obvious use of IPSODE is in the modelling of the water pollution problem described in section 6.5. To recall, water flows into a tank with a pollutant at a prescribed level which varies. The

```
         ** BUGSY **
         T                    CONC                  BUGS
  00.0000E-01           10.0000E-04           10.0000E-02
  50.0000E+00           58.5446E-05           10.1293E-02
  10.0000E+01           34.2439E-05           10.0539E-02
  15.0000E+01           20.1701E-05           09.8618E-02
  20.0000E+01           12.0676E-05           09.6074E-02
  25.0000E+01           09.7284E-05           09.3305E-02
  30.0000E+01           08.4141E-05           09.0513E-02
  35.0000E+01           07.6972E-05           08.7750E-02
  40.0000E+01           07.3336E-05           08.5043E-02
  45.0000E+01           07.1821E-05           08.2407E-02
  50.0000E+01           07.1605E-05           07.9849E-02
  55.0000E+01           07.2198E-05           07.7371E-02
  60.0000E+01           07.3309E-05           07.4974E-02
  65.0000E+01           07.4760E-05           07.2657E-02
  70.0000E+01           07.6442E-05           07.0418E-02
  75.0000E+01           07.8287E-05           06.8257E-02
  80.0000E+01           08.0254E-05           06.6169E-02
  85.0000E+01           08.2314E-05           06.4154E-02
  90.0000E+01           08.4451E-05           62.2083E-03
  95.0000E+01           08.6653E-05           60.3298E-03
  10.0000E+02           08.8912E-05           58.5164E-03
  Would you like to extend your results?Y
  What is the new value of B
1500
  10.5000E+02           09.1223E-05           56.7656E-03
  11.0000E+02           09.3584E-05           55.0753E-03
  11.5000E+02           09.5991E-05           53.4434E-03
  12.0000E+02           09.8443E-05           51.8678E-03
  12.5000E+02           10.0938E-05           50.3465E-03
  13.0000E+02           10.3476E-05           48.8775E-03
  13.5000E+02           10.6056E-05           47.4592E-03
  14.0000E+02           10.8677E-05           46.0895E-03
  14.5000E+02           11.1339E-05           44.7668E-03
  15.0000E+02           11.4041E-05           43.4895E-03
  Would you like to extend your results?N
```

Fig. 7.4 – Print-out of the solution to the pollution problem.

tank is well mixed and the problem is to evaluate a minimum tank size which ensures that the pollutant level in the outflow is below a specified level. In the example to be shown here, IPSODE was used to represent flow through a tank of volume $16,097 \, m^3$ (see section 6.5), when the inflowing pollutant concentration changed from 0.001 to 0.01. The input for IPSODE and SECOND is shown in Figs 7.2 and 7.3. Figure 7.3 shows a student attempt at input for IPSODE and how he misunderstood the meaning of dependent variable and said there were six instead of two. The student realised his mistake when asked to give them names and resolved the situation by typing CONTROL + C and changing his previous input. Figure 7.1

```
Would you like a graph of your results?Y
At which value of T do you want to start your graph
0
What is the final value of T for the graph
1500
Do you want all your variables plotted?Y

** BUGSY **
CONC  0.72E-04      2.57E-04      4.43E-04      6.29E-04      8.14E-04      1.00E-
BUGS  4.35E-02      5.51E-02      0.67E-01      0.78E-01      0.90E-01      1.01E-
*********!**********!**********!**********!**********!**********!*
0.00E+00:                                                          BC
5.00E+01:                                      C                      B
1.00E+02:                          C                                  B
1.50E+02:       C                                              B
2.00E+02:   C                                               B
2.50E+02:  C                                             B
3.00E+02:C                                           B
3.50E+02:C                                        B
4.00E+02:C                                     B
4.50E+02:C                                  B
5.00E+02:C                               B
5.50E+02:C                             B
6.00E+02:C                          B
6.50E+02:C                        B
7.00E+02:C                     B
7.50E+02:C                   B
8.00E+02:C                 B
8.50E+02:C               B
9.00E+02:C             B
9.50E+02: C          B
1.00E+03: C        B
1.05E+03: C      B
1.10E+03: C      B
1.15E+03: C    B
1.20E+03: C    B
1.25E+03:   C    B
1.30E+03:   C  B
1.35E+03:   C B
1.40E+03:   CB
1.45E+03: BC
1.50E+03:B C
*********!**********!**********!**********!**********!**********!*
```

Fig. 7.5 – Graphical output from the pollution problem at the terminal.

shows a complete input for IPSODE without any mistakes. Figure 7.2 shows input for SECOND. The student chose graphical output and was then asked to estimate ranges for the axes. It is probably better to ask for printout first and then follow by a graph, for in this way the user knows what the range of values the variables have. This approach avoids a graph which is mostly blank with all the variables in one corner! A typical printout is shown in Fig. 7.4 and graphical output in Fig. 7.5. These results show that although the pollutant level falls below the legal limit, there is a substantial time period where c is greater than 5.10^{-4} (i.e. up to 10,000 hours).

7.5 DISCRETE SIMULATION

Many dynamic systems or aspects of societal behaviour are amenable to mathematical modelling as queueing problems (e.g. traffic flow on highways or at intersections, planes landing and taking off at airports, customers in a bank or at a store, etc.). However, the majority of students who are non-specialist mathematicians find the more detailed aspects of the queueing theory required to cope with the above systems to be far too difficult if not impossible! Although a number of special-purpose programming languages have been developed primarily, for discrete simulation (i.e. employing the applications of queueing theory), their use generally requires a significant amount of effort and computational experitse.

The ideas of discrete event simulation may be communicated quite simply. There are a wide variety of processes that may be easily recognised and classified as discrete (e.g. ships arriving, docking and departing from port, cars being assembled on a production line, etc.). Such systems may be characterised in terms of customers, queues of customers and servers. Both the customers and servers may be animate or inanimate. Essentially, the simplest discrete system consists of one service point which takes a prescribed time to process each customer. In turn, customers form a queue to be served at the service point (cf. Fig. 7.6). In representing real systems the customers may have a number of tasks, there may be constraints on the queue length or customer waiting time together with some features of the server (e.g. allowance for breaks). Of course, there may be different types of customers, arriving in a prescribed or a random fashion together with a number of servers. The general philosophy of discrete event simulation is described in great detail in a number of texts.

However, it is quite feasible to learn the elements of discrete event simulation by formulating a problem in such a way as to solve it using APHIDS.

```
OLD APHIDS
RUN
  Do you want an description of package and example?N
  Do you want help with your input?Y

  Don't worry if you make a mistake;
  you will have a chance of changing any input section
  If you cannot fit your problem to this design,see the lecturer conce

  Give a descriptive name for the system    SURGERY
  Give a descriptive name for a server      DOCTOR
  Give a descriptive name for a customer     PATIENT
  What is the maximum number of DOCTORS possible?7
  What is the maximum number of PATIENTS possible
  in the SURGERY?100

  How many types of DOCTORS are there?1
  How many types of PATIENTS are there?5
  Give a descriptive name for type  1 of PATIENT?BABY
  Give a descriptive name for type  2 of PATIENT?M/OAP
  Give a descriptive name for type  3 of PATIENT?F/OAP
  Give a descriptive name for type  4 of PATIENT?M/OTHER
  Give a descriptive name for type  5 of PATIENT?F/OTHER

  These are the queueing choices available for you
  ---------- ----------
  1. Each DOCTOR has a queue attached and PATIENTS on arrival
  join the shortest queue at the back.
  2. Each DOCTOR has a queue attached and PATIENTS on arrival are
  assigned in rotation.
  3. PATIENTS join a single queue and an idle DOCTOR takes the
  front of the queue.
  Which one do you want  2

  Is it possible for a DOCTOR to have a short break
  and leave PATIENTS waiting in the queue?N

  This section describes possible actions of a PATIENT
  --- --- --- ---
  A PATIENT enters the SURGERY.It is assumed that when he finishes
  a task,he either leaves or rejoins a queue at the back for his
  next task.
  How many tasks can a BABY-PATIENT have?1
  How many tasks can a M/OAP-PATIENT have?1
  How many tasks can a F/OAP-PATIENT have?1
  How many tasks can a M/OTHER-PATIENT have?1
  How many tasks can a F/OTHER-PATIENT have?1

  Can a PATIENT on entry to the SURGERY decide to leave immediately?N
  Is it possible to wait a time and leave without being  served?N
```

These are the time units available.You are allowed to work
with only one ;make sure your data is consistent .
1 .SECONDS   2 .MINUTES   3 .HOURS   4 .DAYS   5 .WEEKS   6 .MONTHS   7 .YEARS
Please type a number between 1 and 7 2
Interarrival time for BABYPATIENT;do you want a CONSTANT?Y
Interarrival time for M/OAPPATIENT;do you want a CONSTANT?Y
Interarrival time for F/OAPPATIENT;do you want a CONSTANT?Y
Interarrival time for M/OTHERPATIENT;do you want a CONSTANT?Y
Interarrival time for F/OTHERPATIENT;do you want a CONSTANT?Y
Do the interarrival parameters vary with time during the
simulation period?Y
Service times of a BABYPATIENT;do you want a CONSTANT?Y
Service times of a M/OAPPATIENT;do you want a CONSTANT?Y
Service times of a F/OAPPATIENT;do you want a CONSTANT?Y
Service times of a M/OTHERPATIENT;do you want a CONSTANT?Y
Service times of a F/OTHERPATIENT;do you want a CONSTANT?Y

Starting criteria for your model.
———————— ————————
At the beginning of the runtime for the model,which of the following applies
(1) The DOCTORS are idle and the first PATIENT is generated;
i.e the queues are gradually built up.
(2)The DOCTORS are idle but there is a specified number of
PATIENTS already waiting for service.
Please type either 1 or 2 2
Do you want your simulation period to start
(1) At the beginning of the runtime
(2) After a certain time has elapsed
(3) After a specified number of PATIENTS have already arrived
to the SURGERY since the start of the runtime
(4) After a specified number of PATIENTS have already been served
Please type 1,2,3 or 4 1
Stopping criteria for your model.
———————— ————————
Do you want to stop the simulation
(1) After a certain time interval
(2) After a certain number of PATIENTS
Please type either 1 or 2 1
At the closedown do you want to
(1) Stop the simulation and take no account of services underway
(2) Complete current services but refuse to serve waiting PATIENTS
(3) Complete current services and those waiting but stop arrivals
Please type either 1 or 2 or 3 3

Please answer Y or N to each question.Are you interested in:-
Idle time of each DOCTOR?Y
Busy time of each DOCTOR?Y
Number of PATIENTS served per DOCTOR?Y
Waiting times for the PATIENTS?Y
Service times for the PATIENTS?Y
Number of PATIENTS arrived and served?Y
A resume of your model in a file?N

Do you want to change any section?N
Hang on a few seconds while CROSS is being constructed.
Information has been compressed and put into CROSS
Type OLD CROSS,then JS 900LIBA*STICKS and RUN.
D   @   9999

Fig. 7.6 − An example of using APHIDS to construct CROSS.

7.6   APHIDS – A PROGRAM TO HELP INTERACTIVELY IN DISCRETE SIMULATION

APHIDS was specifically designed to enable the 'novice' modeller to develop and carry out discrete simulation tasks without requiring an extensive knowledge of either queueing theory or computer programming. APHIDS is designed in a similar manner to IPSODE insofar as it consists of two sub-programs: APHIDS which constructs CROSS (a Complete Resume Of Simulation System) and STICKS (Simulation TICKS). As for the continuous simulation package, the general description of the problem is constructed via APHIDS whilst specific parameter values are input in STICKS. The two programs are combined in coupled form thus reducing storage space. The problem need only be defined once in CROSS and then many runs can be made with a variety of parameter sets and/or many students can make a run simultaneously.

APHIDS is based on two classes of objects, CUSTOMERS and SERVERS. There can be more than one type of each; for instance, in the example we discuss later, there are five types of customer and one type of server. Each customer can also have several tasks to perform. Several types of customer behaviour are possible. A customer may be discouraged and leave, either because the queue is long or if he is in a queue more than a certain length of time. He may also join the shortest queue (as in a shop) or be assigned to a server in rotation (as in a doctor's surgery) or join a single queue (as at a taxi rank). Also the server may wish to take a break from time to time. The above options are all catered for an their model description is illustrated in Fig. 7.7.

```
OLD CROSS
JS STICKB
RUN
  Give a starting value for a random number seed
876

Please remember to use MINUTES for your time units

How many DOCTORS are there?5

How many PATIENTS are waiting at the start of the runtime?30
                              of type BABY?5
                              of type M/OAP?5
                              of type F/OAP?5
                              of type M/OTHER?7
                              of type F/OTHER?8
```

How long in MINUTES is your simulation period?180
In this model, 30 PATIENTS are waiting at the start of the
runtime and after 0 MINUTES,the simulation starts
The simulation period lasts 180 MINUTES
Please confirm?Y

BABYPATIENT.Service times using CONSTANT
Give the value of your constant 6
M/OAPPATIENT.Service times using CONSTANT
Give the value of your constant 5
F/OAPPATIENT.Service times using CONSTANT
Give the value of your constant 6
M/OTHERPATIENT.Service times using CONSTANT
Give the value of your constant 4
F/OTHERPATIENT.Service times using CONSTANT
Give the value of your constant 5

Interarrival times of BABYPATIENT come from CONSTANT
Give the value of your constant
                                6
Interarrival times of M/OAPPATIENT come from CONSTANT
Give the value of your constant
                                3
Interarrival times of F/OAPPATIENT come from CONSTANT
Give the value of your constant 3

Interarrival times of M/OTHERPATIENT come from CONSTANT
Give the value of your constant .8
Interarrival times of F/OTHERPATIENT come from CONSTANT
Give the value of your constant
                                .8
The parameters for interarrival times may change
After how many more MINUTES do the parameters change?90
Have you input interarrival times correctly?Y

For the graph of dynamic queue sizes,what is the time
interval in MINUTES?30

The results for waiting times and service times
of PATIENTS are presented in either (1) a diagram or (2) a table
Which one do you want.Please type either 1 or 2 1

Do you want to change any other section?N

Give new parameters for the interarrival times. Time is now 90
Interarrival times of BABYPATIENT come from CONSTANT
Give the value of your constant

Interarrival times of M/OAPPATIENT come from CONSTANT
Give the value of your constant

Interarrival times of F/OAPPATIENT come from CONSTANT
Give the value of your constant

Interarrival times of M/OTHERPATIENT come from CONSTANT
Give the value of your constant

Interarrival times of F/OTHERPATIENT come from CONSTANT
Give the value of your constant

After how many more MINUTES do the parameters change?90
Have you input interarrival times correctly?Y

Fig. 7.7 – An example of using STICKS.

The mathematical description of the interarrival times and service times is extremely important and many options are available (i.e. constant, uniform, negative exponential, Poisson, Normal or Erlang). Depending on the mathematical ability and experience of the users the model can be made very sophisticated. However, novice modeller are advised to start with constant interarrival and service times until a feel for the program and the process of discrete simulation is obtained. There are also various options available for stopping and starting the simulation, i.e. whether service stops immediately or doors are closed and all the people in the queue are served, etc.

The various values of the system parameters are specified in STICKS (cf. Fig. 7.8) and the simulation results are then produced in terms of a status report and a plot of dynamic queue sizes throughout the run (e.g. Fig. 7.7).

```
DYNAMIC QUEUE SIZES
---- --- ---- --- --- --- --- ---


              TIME IS 150 MINUTES
QUEUE 1 ****************
QUEUE 2 ***********
QUEUE 3 ********
QUEUE 4 ************
QUEUE 5 *****************
              TIME IS 180 MINUTES
QUEUE 1 **************
QUEUE 2 **********
QUEUE 3 *********
QUEUE 4 **********
QUEUE 5 ***********
++++++++++++++++++++++++++++++++++++++++++++++++++++++++++++++++++++f

        REPORT  PRELIMINARY TO SIMULATION  OF SURGERY
        --- --- ---

    Total number of DOCTORS in the SURGERY                = 5

    DOCTOR    state at      idletime    busytime     served
              present                                tasks
       1      idle            14          267          58
       2      idle             0          281          58
       3      idle             6          275          58
       4      idle             6          275          58
       5      idle            14          267          57


    PATIENT         NUMBERS
               arrived        served      left immed    now present
       BABY       170           32           138            0
       M/OAP      185           27           158            0
       F/OAP      185           17           168            0
       M/OTHER    269          129           140            0
       F/OTHER    270           84           186            0

       RUNTIME =   281
++++++++++++++++++++++++++++++++++++++++++++++++++++++++++++++++++++++
```

Fig. 7.8 – Output from the APHIDS discrete simulation package.

The principles used in the development of APHIDS were similar to those used in IPSODE, so that errors are trapped wherever possible and helpful diagnostics given. Again APHIDS has been made user-proof in the sense that the package will not abort on the input of incorrect data, but will try and help the user to follow the procedures in the correct manner with acceptable (though not necessarily correct) data.

Consider the following example:

Five doctors share a group practice and must decide if they are to have an extra partner for evening surgeries or change their hours. Surgery hours start at 5.30 p.m. (but doors open at 5.00 p.m.) and shut at 8.30 p.m. The doctors continue until they finish their allocation of patients who are assigned in rotation as they enter the waiting room. The arrival rates and service times are known for each type of patient. There are five types of patient: babies, male and female old age pensioners, other males and females. There is also a maximum number of doctors and patients allowed in the surgery at any one time.

This problem may be prepared for APHIDS in the following manner: The SERVER is a DOCTOR, the CUSTOMER is a PATIENT and the SYSTEM is a SURGERY. The five patient types are BABY, M/OAP, F/OAP. M/OTHER, F/OTHER. The maximum number of doctors is six and the maximum number of patients is 100. The interarrival times and service times are given by the following tables:

p.m.	Monday	T/W/Th	Friday
Interarrival times in minutes for baby			
5.00–5.30	6	10	5
5.30–7.00	9	12	10
7.00–8.00	18	20	15
Interarrival times in minutes for OAP			
5.00–5.30	3	5	3
5.30–7.00	1.5	2	1.5
7.00–8.00	3	8	4
Interarrival times in minutes for others			
5.00–5.30	2	3	1
5.30–7.00	0.4	0.9	0.6
7.00–8.30	2	3	1

The service times are constant:

	Male	Female
Baby	6	6
OAP	5	6
Others	4	5

The input is shown in Figs 7.7 for CROSS and 7.8 for STICKS. Figure 7.7 demonstrates how the general description of the customers and servers is prescribed for the simulation. Interarrival and service times are to be taken from the tables above and for this example are constants. In Fig. 7.8, it was decided to start the simulation at 5.30 but allow for patients who have been arriving since 5.00 p.m. Decisions on how to stop the simulation and the choice of the output features to be monitored are also made. The general description of the problem is then complete and all the relevant information is stored in a file called CROSS. In Fig. 7.8 it is decided to run the simulation for 3 hours (i.e. 5.30 p.m − 8.30 p.m.) and input the arrival and service times for the first 90 minutes. It is taken that there is 0.5 probability that a patient could be male or female and the interarrival times are adjusted accordingly. Fig. 7.7 shows how results can be presented and a report is given after the first 90 minutes. By answering NO to the question 'Do you want to finish?' the interarrival and service times for 7−8.30 can be entered and another report obtained.

7.7 CONCLUSION

The development of cheap, flexible and user-friendly interactive computing systems together with robust numerical techniques has made the advantages of mathematical modelling available to many areas of professional life in modern society. To be able to use models sensibly, putative professionals (i.e. students) must learn something of the art of mathematical modelling even though they are likely to possess fairly rudimentary mathematical, numerical and computational skills.

In this chapter, the tools required to teach essentially non-numerate (or at best non-specialist) students the art of mathematical modelling are briefly described. These educational tools consist of two interactive computer simulation packages – one each for continuous and discrete systems. The packages have been used with some success in a number of undergraduate courses on mathematical modelling at Sunderland Polytechnic over the last few years. This success has encouraged a number of other course committees to request mathematical modelling courses using these simulation tools. Whilst it is obvious to suggest that specially written simulation packages for students makes sense, it is no easy task to write them in a suitable fashion. Although IPSODE and APHIDS are working satisfactorily they may not have taken the optimum approach. As further operational experience is gathered with succeeding generations of students it is hoped to improve and fine-tune these aids to modelling and simulation.

CHAPTER 8

Conclusions

8.1 WHAT USE IS THE COURSE?

Since the mid-1960s the role of the mathematician in industry or commerce has evolved quite substantially. Gone is the organisation whereby mathematicians were cloistered together attempting to generate solutions to equations or make statistical sense of data that arrived 'post box' style. Today mathematicians are generally part of a project team – as such they develop expertise about the process of system being analysed rather than act merely as solvers of mathematical equations. Their involvement covers:

(1) rationalisation of a perceived problem into a 'real' one;
(2) sorting of relevant data from a mass of irrelevant information and the formulation of an appropriate mathematical model;
(3) solution of the resulting equations utilising computer technology;
(4) judgement about the adequacy, limitations and utitity of the model as a representation of the perceived 'reality';
(5) use of the model in detailed process or system analysis;
(6) presentation of results using computer graphics, preparation of readable reports and oral presentations to convince the sponsor of the efficacy of the implications of the analysis;
(7) participation in the implementation of recommendations.

As reliable and robust numerical software become available for various classes of problems (e.g. fluid flow, stress analysis, networks, etc.), proportionately less effort goes into generating a numerical solution to the set of model equations. It is clear – as was affirmed on the first page of this book – that mathematical modelling involves rather more than the ability to solve sets of equations.

Conventional degree courses in mathematics tend to concentrate upon *analysis* in its abstract sense. As a result, great effort goes into analysing specified classes of equations — deciding whether solutions exist and commenting upon their uniqueness. By comparison, very little effort traditionally goes into investigating the context in which the set of equations were derived — even in applied mathematics. No, the effort tends to be placed upon the mathematical analysis. Notice, the emphasis is on the analysis of the set of (often generalised) mathematical equations rather than the process or system being investigated. However, for mathematicians working in industry or commerce, the opposite is true: the objective is to provide (at the least) insight into the system or process being analysed — the mathematics and associated computing activity are merely the tools. Viewed in this light it is clear that, unless undergraduates are exposed to the activities of mathematical modelling in their degree course, then as graduates they will be ill-prepared for careers in industry or commerce.

The objective of the material in this book is to provide mathematicians with some exposure to the activities involved in modelling real 'life-like' processes and systems. Obviously, within such a restricted framework as an undergraduate course, one cannot hope to address each of the activities of mathematical modelling in any depth. However, if approached in the right way the material in this book can provide reasonable experience of items (1) to (4) together with some exposure to items (5) and (6) of the above list. Therefore, a diligent student should certainly acquire some of the important skills needed to be able to analyse systems or processes via mathematical modelling. That, simply, is the potential utility of the book and the course from which it derives.

8.2 SOURCES OF FURTHER MATERIAL
AND ACADEMIC GROUPS

In this short book we have tried to demonstrate that students can learn the essentials of mathematical modelling if they are placed in the correct environment with an appropriate progression of problems. It is worth noting that the activity of 'teaching' mathematical modelling is not small. Most polytechnics and a good few universities now have a block on mathematical modelling in at least one of the years of their mathematics degrees, and, increasingly, in other courses too.

There is a growing fraternity of academics involved in mathematical modelling now. Two journals cater to a large extent for the needs of educational modelling:

- IMA journal on *Teaching Mathematics and Its Applications*, edited by Professor D. Burghes.
- *International Journal of Mathematical Education in Science and Technology*, edited by Professor Bajpai.

In addition, there are a number of books with compilations of modelling scenarios sutiable for undergraduate level:

- D. J. G. James and J. J. McDonald (eds), *Case studies in Mathematical Modelling*, Stanley Thornes (1981).
- D. N. Burghes, I. D. Huntley and J. J. McDonald, *Applying Mathematics*, Ellis Horwood (1982).
- D. N. Burghes and M. S. Borrie, *Modelling with Differential Equations*, Ellis Horwood (1981).
- D. N. Burghes and A. D. Wood, *Mathematical Models in the Social, Management and Life Sciences*, Ellis Horwood (1979).

Some problems at a more advanced level are provided by

- R. Bradley, R. D. Gibson and M. Cross (eds), *Case Studies in Mathematical Modelling*, Pentech Press (1981).

Also, the following books contain some useful descriptions of models rather than student-oriented scenarios *per se*

- J. Andrews and R. R. Mclone, *Mathematical Modelling*, Butterworth (1976).

There are also three international journals devoted to the presentation and discussion of mathematical models developed by research workers

- *Applied Mathematical Modelling*, edited by Mark Cross.
- *International Journal of Mathematical Modelling*, edited by Xavier Avula.
- *International Journal of Mathematics and Computer Simulation*, edited by R. Vichnevesky.

Although there is no formal organisation of those involved in mathematical modelling, there are a number of informal groups:

(1) The Spode Group, which is specifically aimed at developing class material is contactable through

  – Dr I. D. Huntley, Sheffield City Polytechnic.
  – Dr John Berry, The Open University.

(2) The Mathematical Education Group at the University of Nottingham, where the contact is Dr H. Burkhardt.

(3) The Mathematical Education Group at Exeter University, which is headed by Professor D. N. Burghes.

In the North East a group have collaborated to form POLYMODEL, which encourages modelling at a postgraduate level. However, there have been substantial spin-offs and they can be contacted through A. O. Moscardini at Sunderland Polytechnic.

We have only mentioned groups that we are well aware of, there are yet more in North America, Australia and New Zealand.

8.3 HOW DOES THE COURSE BLEND WITH EXISTING MATHEMATICAL DEGREE SCHEMES?

From the brief comparison between the existing emphasis of academic mathematics, as represented in its degree courses, and the needs of a modeller, it is clear that there is a substantial difference. This difference is manifested in the philosophical approach of mathematical modellers and conventional academic mathematicians. If a mathematical modelling course is 'dropped' into an existing conventional degree scheme, it will appear as out of place, both to the students (who will, at least, perceive the implications of the different philosophical approach) and to the other academic mathematicians teaching on the course. The latter tend to greet modelling components with a great deal of scepticism and, sometimes, open opposition. There are cries of, 'It's not mathematics but engineering', 'But they don't learn any new mathematics', and 'What has happened to rigour?'. Add to this the fact that (1) there are difficulties in assessing modelling assignments and that (2) most courses contain very little of practical computing science, and it is clear that modelling courses fit uncomfortably into conventional mathematics degree schemes.

8.4   A NEW APPROACH TO MATHEMATICS DEGREE COURSES

What is the fate of mathematical modelling components in mathematics degree courses? Probably one of two: either the modelling dies when the interested academic leaves or the needs of modelling begin to impinge upon other aspects of the degree course.

Essentially, practising mathematicians (as opposed to most academics) work in the context of the computer technology available to them. Depending upon the hardware and software available, mathematicians can develop and use models at one of a number of levels of complexity to analyse processes or systems. The more the mathematician understands about

(i)   the process under investigation,
(ii) the implications of computing machinery available,
(iii) practical numerical and mathematical techniques,

the better he is equipped to build the most appropriate model, in the circumstances, to analyse the process under consideration.

It is clear from the above paragraphs that it is vital that, if mathematics graduates are to be technically effective in industry or commerce, then their degree courses must contain a substantial amount of 'relevant' computing science. It is important that such graduates can develop good well-structured software, with expertise in data processing (particularly database concepts) and information display techniques (e.g. graphics). It is vital that the teaching of numerical and statistical techniques never strays far from the context of computer implementation. In the same way, numerical and statistical *analysis* should take cognisance of the computer technology that will be used to implement the schemes. In other words, somewhat less abstraction and rather more pragmatism is called for. Graduates need to be able to make judgements about the efficacy of various techniques, and to do so they need pragmatic criteria which are at the same time improving their understanding of the set of equations which comprise the model.

Some institutions are finding it useful to teach mathematical and numerical analysis together as a single course. If this course is twinned with another on mathematical software development, the content of the former course becomes less than contentious, but the combination is both academically challenging and relevant to the needs of our technological society. Of course, it is useful to have some of what is conventionally called pure mathematics to underpin (perhaps, more appropriately, to provide some mathematical background and context to) the more practical computational mathematics; but the proportion is substantially reduced from conventional degrees. Similar approaches could be taken with algebra-related subjects and statistics. In this way, the whole subject becomes more integrated and the material more relevant to the needs of the graduate in industry or commerce. Graduates of such courses may then enter industry with at least a nodding acquaintance with the origins of their mathematics (i.e. rigour, uniqueness, existence, etc.). However, most importantly they will have an ability to set about formulating problems in the real world as a set of mathematical equations, developing a practical solution to those equations and using the resulting model to provide insight into the problem under investigation.

Index